Praise for

Tinsel

"A delicately calibrated combination of rigorous reporting, observational humor, and old-fashioned empathy, *Tinsel* is the book that saved Christmas for this curmudgeon."
— Laura Miller, *Salon*

"A study of Christmas excess as an exercise in American anthropology . . . [Stuever] manages to find the heart in his characters' obsessive consumerism [and] the somewhat jolly realization that no matter how prepackaged and homogenized the holiday has become, something about it remains inviolably personal."
— *Entertainment Weekly*

"The book you need to read to get ready for the season is Hank Stuever's lively *Tinsel* . . . Stuever is both a magnificent prose stylist and a compelling storyteller, and his richly detailed reportage rings true . . . The book doesn't judge; it reveals."
— Rod Dreher, *Dallas Morning News*

"Amazingly reported." — *New York*

"A stylishly written and often delightful book that aims to capture all the things that Christmas is about — family, values, religion, ritual, celebration, kitsch . . . [Stuever] is wry and witty and astute."
— *Washington Monthly*

"Stuever's clear-eyed examination of America in holiday-orgy-mode is energetic, acerbic, and informative."
— *The Stranger*

"Stuever unwraps both appalling consumerism and genuine holiday spirit — sometimes in the same package — and treats the people he writes about with respect and affection, even when they're doing things he can't quite believe."
— *St. Petersburg Times*

"What emerges [in *Tinsel*] is much more than the contest between sacred and secular. It's how the happiest time of year can also be melancholy and how holiday rituals collide with popular culture." — *The Oklahoman*

"Marvelously written and sharply observed. You will definitely laugh; you will probably learn; you might get angry . . . Some reporters go to the halls of government to take the pulse of the nation; [Stuever] goes to Bed Bath & Beyond. The man can see the secret life of America in a Slurpee cup."
— Patrick Beach, *Austin American-Statesman*

"Reading *Tinsel* is a nice antidote to the blizzard of obligations, expectations, and traditions that bury us at the end of each year."
— *Cleveland Plain Dealer*

"The ultimate holiday drop-in." —*Redbook*

"Scrupulously observed . . . *Tinsel* is not just the obligatory snapshot of America with Santa, it is a meticulously detailed portrait of a society that remains true to its hunter/gatherer/capitalist roots." — *San Antonio Express-News*

"Immensely entertaining . . . A fast-paced examination of what December 25th has done to Americans, and vice-versa."
— *Buffalo News*

Tinsel

Books by Hank Stuever

Off Ramp:
Adventures and Heartache in
the American Elsewhere

Tinsel:
A Search for America's
Christmas Present

Tinsel

A SEARCH
FOR AMERICA'S
CHRISTMAS PRESENT

Hank Stuever

Mariner Books
Houghton Mifflin Harcourt
BOSTON • NEW YORK

First Mariner Books edition 2010

Copyright © 2009 by Hank Stuever

For information about permission to reproduce
selections from this book, write to Permissions,
Houghton Mifflin Harcourt Publishing Company,
215 Park Avenue South, New York, New York 10003.

www.hmhbooks.com

Library of Congress Cataloging-in-Publication Data
Stuever, Hank.
Tinsel : a search for America's Christmas
present / Hank Stuever.
p. cm.
ISBN 978-0-547-13465-9
ISBN 978-0-547-39546-5 (pbk.)
1. Frisco (Tex.)—Social life and customs—Anecdotes.
2. Stuever, Hank— Homes and haunts.   3. Christmas—Texas—Frisco.
4. Frisco (Tex.)—Biography— Anecdotes.   I. Title.
F394.F935S78 2009
976.4'556   [B 2 22]   2009013746

Book design by Melissa Lotfy

Printed in the United States of America

DOC 10 9 8 7 6 5 4 3 2 1

Lines from *A Charlie Brown Christmas* by Charles M. Schulz
and Introduction by Lee Mendelson. Copyright © 2000 by
Lee Mendelson. Reprinted by permission of HarperCollins
Publishers

Lyrics from "Mary, Did You Know?" written by Mark
Lowry, Buddy Greene. Published by Word Music, Inc.,
Rufus Music (ASCAP). Used by permission.

*This book is dedicated to
Laura McCall Froelich*

*And to my mother, Joann,
for Christmases past
(especially 1982)*

*And to Michael Wichita,
for Christmases future.*

Those about me, from childhood on, had sought love. I heard and saw them. I saw them rise and fall on that wave. I closely overheard and sharply overlooked their joy and grief. I worked from memory and example.

— LOUISE BOGAN, *Journey Around My Room*

"Thanks for the Christmas card you sent me, Violet."

"I didn't send you a Christmas card, Charlie Brown."

"Don't you know sarcasm when you hear it?"

—*A Charlie Brown Christmas*

Contents

Crèche

Half Off

Baby, Please Come Home

A NOTE TO THE READER

The people and places you'll encounter in this book appear with their real names, except in a few cases:

Tammie Parnell and I agreed that I would change or omit the names of her Christmas decorating clients and blur some details about their homes. I also changed the names of Tammie's friends.

In scenes from the mall, I changed the names of the Israelis working at the Snow Powder cart because of their iffy visa status. They disappeared before I could learn their full stories.

Caroll Cavazos's older daughter, Michelle, asked that I leave out her and her husband's surname.

Jeff Trykoski asked me to leave out the name of the Chinese factory where his LED lights are made.

Best Buy

(A PROLOGUE)

BEFORE THE BLACK FRIDAY DAWN, the sky is still a mix of dark blue and the sick sodium-vapor saffron of the suburban night. I park by the Beijing Chinese Super Buffet and walk across the lot to Best Buy, where hundreds of people — some in their twelfth or thirteenth hour of standing in line — await the day-after-Thanksgiving doorbuster sale. Best Buy will open at 5 A.M. The shoppers are wrapped in their fleecies, hoodies, and wubbies. They have their grande lattes and their Krispy Kremes. Some pitched tents and now have their butts planted on portable reclining chairs that were purchased for the specific act of waiting around, waiting all over America, waiting as they've learned to do when Harry Potter novels are released, or when new generations of video game systems come out, or when reality TV producers hold auditions. The line wraps around the big box. A news helicopter flies overhead to show the world itself at the beginning of another holiday season, and the theme never changes: *See what it's come to.* Everyone looks up at the sky. Christmas is at our throats again.

This is the Centre at Preston Ridge, a mammoth retail strip mall, one of several "power centers" (as the real-estate guys call them) in Frisco, Texas, a boomtown outside Dallas. I walk past chain restaurants, box stores, and boutiques — including Fetal

Fotos, a place to get ultrasounds-on-the-go, a Baby Jesus moment right there. So smitten were the developers with their site's pioneer lore that they incorporated it into the design: Bronze sculptures of cattle roam the landscaped berms between rows and rows of parking spaces. Three limestone obelisks bear never-read plaques telling the story of the Shawnee Trail, which ran through here a century and a half ago. It's the mythic saga of the millions of cattle driven through this very land and the difficulties faced by homesteading settlers, as if gently chastising happy consumers, *Just you think about that hardscrabble life next time you're wandering through Old Navy.* Over another short hill, by the Hampton Inn, more bronze sculptures of longhorn steers are on a permanent stampede toward the Target and the TJ Maxx.

Several of the Best Buy employees (corporate calls them "blue shirts") are imploring those of us in the crowd to stay calm. A couple of cops are here, too.

I'm hanging back with a mother I've just met, who's in her late forties. She has straight, shoulder-length brown hair, a nice, nervous laugh, and a look of determination in her eyes. Her daughter, who is ten, is wearing a fuchsia shearling jacket, her long brown hair pulled back in a ponytail, and she twirls around and around with caffeinated anticipation, talking and talking. The girl tells me she wants a pink iPod Nano for Christmas. She tells me she's going to be Lucy in her school's production of *A Charlie Brown Christmas.* She sees a Hummer H3 parked nearby and wishes her family had one. "But we drive a Taurus," she says. The girl tells me her name is Marissa, and I write it down. ("Why do you write everything down?" she asks.)

The mother tells me her name: Caroll, with one *r* and two *l*'s. Caroll tells me what she's here to buy today: a computer for her own mother, a washer and dryer for her older daughter and son-in-law, and a laptop for her son. I ask Caroll if it's going to be a big Christmas for her family this year.

"Well, I don't know," she says. "What's 'big'?"

Christmas is the single largest event in American communal life, intersecting with every aspect of religion, culture, commerce, and politics. From mid-November to New Year's Eve 2006, shoppers spent almost half a trillion dollars on gifts, which is more than we spend on almost anything else as a people, including the annual bill at that time for ongoing wars in Iraq and Afghanistan. It's a staggering sum to consider unless you are the retailers and investors who pay close attention to it each year; for them it is never enough. A steady drummer-boy paradiddle of worry and gloom begins on the September morning when the National Retail Federation, a Washington, D.C., lobbying group, releases its annual holiday sales forecast, with analysts soon chastising the consumer for not spending enough to impress stock investors. Even in healthy, average years, when retail sales increase 4 or 5 percent from the Christmas before, the holiday is usually portrayed as a great letdown.

For those who opt in, Christmas is supposed to exist as a pure moment of bliss and togetherness. We spend more money than we have at Christmas in part to get closer to the simple joy it advertises. Many millions of people find inspiration in Christmas every year — unpacking it just where they stored it last year and basking in its returning glow. The years go by, babies are born, and it's supposed to get better each time.

Except, but, *however*. Because it looms so strong in memory, and because the happiness it represents can be so achingly elusive, Christmas can bring us down like almost nothing else. It is our happiest day and our greatest disappointment all in one, heavily freighted with expectation and the loss of innocence.

Christmas pressures undo Nativity naiveté, delivering the anti-climax as young Virginia discovers (and she will always discover) that there is no such thing as Santa Claus, that her family has myriad emotional issues, and that her December receipts have a way of adding up to a negative balance come January. Virginia, no longer a child, becomes a consumer instead. Many of our TV specials, movies, and songs about Christmas revolve around the act of saving the damaged holiday, rescuing it, returning it to its simpler (if inaccurate) origins. We scream at one another on talk shows about where Christmas has gone wrong.

There was a time I imagined this book as a *Fast Food Nation* or *The Omnivore's Dilemma,* only about ornaments. I wondered if I could be the sort of writer who would, say, journey to China to watch factory workers (oppressed elves?) get paid deplorable wages to trim the threads on freshly manufactured holiday sweaters destined for American department stores, to be worn at ironic "ugly sweater" parties in lofts owned by trendy First-Worlders. I prepared to go that way, called economists and retail analysts, recorded interviews and then transcribed them, took notes and printed out spreadsheets. I had, and still have, the numbers that define Christmas as unambiguous math.

I also considered the literary allure of the sometimes capricious form of memoir: I could rip out my own heart and examine it, to figure out where exactly "the most wonderful time of the year" and a nice heartland boy like me went our separate ways — while keeping it *funny.* Maybe my story could be a poetic analogue to many Americans' own mixed feelings about the holiday. Or what about other writerly gimmicks? I could, for example, make an old-fashioned American road trip to every year-round Christmas decoration store in the land (there are 1,500 of them), until I found my carefully calculated special ending. I could end up in some snowy village next to an outlet mall. I'd

4

trim a tree, gladden a child, and head toward the happy light. Maybe Owen Wilson could play me in the movie version (Owen doing an urbanized, blue-state gay guy journeying into a suburbanized, red-state Christmas). I dropped my popcorn and fled that particular mental multiplex just as the first reel began.

I wanted this story to be about Christmas but also everything else: our weird economy, our modern sense of home, our oft-broken hearts, and our notions of God. The biggies. Where novelists and the makers of romantic holiday comedy movies exaggerate and fictionalize the Christmas of collective memory, I desired something more true, to see Christmas in the high-definition light of the early twenty-first century — starting at the ass-crack of dawn in front of Best Buy with hundreds of shoppers — and not warm myself by the make-believe fireplaces seen in those commercials for diamond necklaces, or trap myself in the black-and-white, Bedford Falls reruns of forever (and never) ago.

After years of sitting it out, doing whatever I could to avoid the holiday's many meanings, rituals, and drudgeries, I set out to do this book because I wanted back in. I wanted to surrender to whatever Christmas has become and study it with as little visible flinching and as much quiet note taking as possible. I went looking for an America living not only on borrowed time, but also on borrowed grace. In the Nativity pageant I've staged here, I cast myself merely as an extra, a Wise Man in a purple velour bathrobe and a cardboard Burger King crown, *following yonder star,* bearing my mother's crystal salad-dressing cruets (my frankincense, my myrrh) on a tasseled living room throw pillow.

This book takes place over three holiday seasons (2006, 2007, and 2008) among three unrelated families who live in a new megaworld north of Dallas, a place that often seemed to have surrendered its truest identity to the shopper within. From my first Christmas there to another (and then to another), certain key aspects of the agreed-upon economic fantasies we all live in came apart: people began losing the easy money that once made them feel so rich, and neighborhoods of pitched-roofed modern castles saw foreclosure rates double, triple, quadruple. Americans became keenly aware, some for the first time in a long time, that their lives had been spent in a suspended sense of entitlement, in a country that buys instead of makes, with money it does not technically have. I did not realize that the Christmas of 2006 would, by the time I finished my reporting, begin to resemble a *belle époque.*

Part of this story is about that twilight moment, when the arrows on the charts had only just begun to point down. If greed was the sin and if economists were the preachers, then these could have been the beginning of the apocalyptic End Times. Yet I went looking for beginnings, birth metaphors, manger scenes. I stepped deliberately into the family-centric, reindeer-sweater-wearing Yule of Baby Jesus, in the newest and most homogeneous America, seeking whatever Christmas was left in me by venturing to a place that appeared to have plenty of Christmas to spare. I looked for consumers who held the firmest faith in God and the dollar and asked if I could follow them through their entire holiday, and then the next year's, and then the next. I embedded with the people whose house has the best lights display, timed to blink on and off to simulcast Trans-Siberian Orchestra and Mariah Carey pop carols. I was a charity angel rushing to buy Bratz dolls and hand-held Nintendo DS players for the Tiny Tims of the day. I helped put up thousand-dollar plastic trees in brand-new McMansions: the twelve-foot faux fir in the entry foyer; the eight-footer in the family room; and the five-footer, with the pink branches and the lavender ornaments, in the baby's room.

In this search (escapade? immersion?) I came upon one word over and over, emblazoned on various plaques, ornaments, and other bric-a-brac. It was at every holiday crafts bazaar I went to, or somewhere in the holiday décor of every house I visited — soldered in pewter, or sewn to Christmas stockings, or decoupaged onto wood. The word was *believe*. Snowmen held signs with it. A team of reindeer pulled it, B-E-L-I-E-V-E, across a front lawn. It was projected in fifteen-foot-tall letters at church Christmas pageants. *Believe*, people kept telling me.

I told them I wasn't that kind of believer.

I told them this is probably not that kind of Christmas book.

I want you to know that.

Caveat emptor, and so forth. (Not valid with other offers. All returns must be accompanied by a receipt.)

"What's 'big'?" Caroll asks.

I love her then and there.

She wonders if maybe this is how memories are made now. Maybe the shopping *is* the memory itself. In her heart, Caroll Cavazos knows shopping is not her reason for the season, but this hasn't seemed to stop anyone else. There are savings here, no doubt, but also the chance to be swept up in something large and thrilling. While it may look absurd tonight on the news, Black Friday can also be seen as a shopaholic's annual Woodstock, or the American version of the running of the bulls at Pamplona. Is Black Friday really any more ridiculous? Is it really any different from the impulse to be part of a rushing mob of humanity? This is how Christmas began, after all, way before the time of Christ (and for centuries after his death), when it was a pagan celebration of the winter solstice. People gathered and danced by fires. The harvests were in and everyone gorged.

Caroll tells me it's one of her favorite days of the year, because it feels like her family is together as one, in some exciting way. It's the rare morning she treats herself to a $4 latte at Starbucks. Hers is the American Christmas retail story told in a very small and personal way, among the box stores and mega-churches. All Caroll ever wants for Christmas is a day when everything's okay. Not flashy, not perfect, and not spending herself into a hole; just a Christmas when she gets "to do for my kids."

The blue shirts push the electronic doors open and people hustle into the store, fanning out toward the flat-screen televisions and DVD players, launching themselves toward games, toward the car stereos, refrigerators, and boxed sets of their favorite old TV shows. Caroll and Marissa step into the whole glorious, messed-up thing.

"Come with us," Marissa says, and I do.

Fake Is Okay Here

✧

(ONLY 52

SHOPPING DAYS

TILL CHRISTMAS!)

1

Target

IN THE SPRING OF 2006 a meteorologist in Washington, D.C., draws for me a red line across a map of the continental United States, below which the odds for a white Christmas grow longer. I set out to be way south of this line, to find people in drought-prone Sunbelt states dreaming of white Christmases that they know will probably never come. (The magic is in the hoping, correct?) I'm searching for some perfect anywhere to set this tale. I have maps and demographic breakdowns. I have store directories to shopping malls unknown to me.

I stick a pushpin on Texas:

Frisco.

The population has grown from about 6,000 people in 1990 (a farm town, mostly) to about 90,000 in the fall of 2006 (a shopping-center town), the same year *Forbes* named Frisco the seventh-fastest-growing exurban city in America.

In a North Texas Tollway Authority promotional video, a developer brags that Frisco now has more footage of chain retail shopping and dining in a single square mile than anywhere else in the Southwest. Almost all of it came after the grand opening of a 1.1-million-square-foot indoor mall called Stonebriar Centre in 2000, meaning that everything Frisco prides itself on — a new, retro-style village-like downtown plaza and City Hall, new

police and fire headquarters, a five-floor main library, a professional league baseball stadium, hockey rink, and soccer arena,
and more than two dozen new schools — is in some way thanks
to sales tax revenues brought in by retail developments. The
rest is built on bond issues and on property taxes collected from
a potentially endless vista of planned subdivisions, where the
average home offers 3,000 square feet of living space and, at the
time, sells for an average price of $219,000. Frisco's average an-
nual household income is $93,000. The average age is thirty-
one. Averagely averaged.

In the late summer I leave my job as a newspaper reporter in
Washington and find a room to rent in a soaring-ceilinged, four-
bedroom brick house (what passes for a "starter home" in these
parts) at the end of a cul-de-sac in the Lakes of Preston Vineyards subdivision in Frisco. A sweet-faced former Marine just
bought the house and lives here with his tanned, lithe, Jennifer Aniston–like girlfriend. They both work in high-tech office
parks in different directions, miles away, with morning commutes so grueling they wake at 5:30 and sneak away, like
thieves.

I spend a lot of my days in the Stonebriar Centre mall, where
mommies in Adidas track pants — their dry, golden hair pony-
tailed through ball caps — push Bugaboo and Peg Perego sport-
utility strollers every afternoon in long, endless laps across the
glossy white travertine floors, past the Coach handbag boutique
and the Aveda beauty store, past the Build-A-Bear Workshop,
past the Nordstrom, past the size-two mannequins, hips slung,
in the display window of Forever 21. I see possibility, fecun-
dity, annunciation, visitation. The angel of the Lord appears any
minute now and he says, *Coming Soon to Stonebriar Centre*. He
says, *Hello, and welcome to Ikea*. And as it was for sweet, obedi-

ent Mary, one can give no reply but *yes*. Lo and behold, beyond Stonebriar, the endless box stores, the vast parking lots, the construction cranes laying virginal freeway overpasses.

Sometimes even the people feel brand-new — in pretty gift wrap. Billboards on newly widened streets advertise Lasik so you can see new, cosmetic veneers so you can smile new; 1-800 numbers extol the miracle of reverse vasectomies, because new things are happening all the time. People smile at me with brilliant white teeth, and before long they are hugging me hello and goodbye. I learn how new and improved pairs of frankenboobs feel as they briefly press against my chest in understanding hugs of welcome.

I work out one morning with the young moms of StrollerFit, who meet every weekday before the stores in Stonebriar Centre open and who in part gave the mall its nickname: Strollerbriar. The StrollerFit instructor lends me her stroller *and* her kid. Our class stretches and runs and crunches and lunges with our strollers; we skip rope in front of our transfixed, saucer-eyed babies while we huff and puff and sing "Itsy Bitsy Spider" to them, double-time. By the second set of plié bends, near the Sears, I start to hurt.

A thousand years ago, the natives mostly passed through this land, using the high, flat ridge as the easiest path to someplace else. The cowboys drove cattle. A collection of homesteads eventually made a town, composed of people who took the federal government up on an 1841 promise of 640 acres per family, free and clear, in exchange for five years of determined effort to stay put and make something out of all that empty Texas. They ranched; they grew cotton. The train tracks came in 1902, and the new depot town was given the same nickname people used when referring to the St. Louis & San Francisco Railroad: *Frisco*.

The geology, the Native Americans, and the frontier past are dealt with quickly enough on the mural painted above Stonebriar Centre's food court.

Eight hundred people a month are now moving here. The talk is always about the burgeoning growth. These newest settlers measure it by what stores and restaurants opened after they got here, on vacant land now filled: only a few years ago, "there were just cows" where the Outback Steakhouse and the Red Lobster and the Bed Bath & Beyond are; just a cotton field where that Kohl's is. "Did you know that where the Mattress Giant is, used to be a whorehouse?" (There was more than one whorehouse, as recently as a decade ago, with come-on names like April's and the Tub Club and the Doll House. Gone now.) There was just the fruit stand on the two-lane road. Just the frozen custard stand, just an old Quonset hut, just a little white church in the distance, just the ramshackle, abandoned farmhouses from the early 1900s, which the fire department started burning down for practice.

When I got here, people say.

When we first bought in this neighborhood, people say.

Where the Home Depot is now.

People have been here six months. They have been here eight years. They have been here about a year and a half. They come and go on the corporate "relo." A few residents have been here all along, and a very few of them remember events such as the catastrophic 1922 fire on old Main Street, or the glory days of the railroad. These original citizens and their descendants can now drive on brand-new parkways and pay visits to classes in just-opened elementary schools that are named after themselves, their parents, or their great-meemaws.

I drive around for weeks in the late summer and early fall and take notes obsessively. (Notes on what, about what, because of what — I can't yet be sure.) Stacked freeways are the prime geographical feature, a favorite Dallas vantage point, and a symbol of achievement. I veer and merge dreamily over and across the great grids that form a region of twelve counties and 6 million people. I take the Lyndon Baines Johnson Freeway (the LBJ) to the Central Expressway to the President George Bush Turnpike (the PGBT) to the Dallas North Tollway. Where the LBJ meets U.S. 75 is an interchange of stacked ramps called the High Five, which is so revered by drivers that the *Dallas Morning News* asked its readers to send in their most stirringly artistic photos of it. Lately a few citizens have been leaping to their deaths off the High Five. I can think of no greater compliment to the builders of intricate freeway spans than to attract both poetry and despair.

At Target, women in full makeup seem to bump into my cart simply for the pleasure of cooing, *Excuse me, I am so sorry* in a delicious twang. People are nice until they aren't. They vote against illegal immigration at every opportunity, only not when they need bathroom tile laid or lawns mowed or Christmas lights hung. They tell me over and over that global warming is a gigantic myth, a product of liberal propaganda. They live in fear of autism, cancer, child predators. We are very pink ribbon in Frisco; we are very yellow rubber bracelet. We are constantly asking one another for prayers. Sixty-eight percent of Texas residents in one poll said they believed the Bible to be the literal word of God. Eight of ten voters here picked George W. Bush for president in 2004, and, when I'm stuck in traffic (on the turnpike named for his father), I notice that many of them have put old Bush/Cheney '04 bumper stickers on new '07 cars and trucks.

It is a place to watch, and be watched. Police officers and mall parking lot security guards can't abide all my inactivity, my stakeout of no one, my meticulous investigation into nothing in particular, and they pull alongside my car, pleasant as can be, asking me to roll my window down: *Can I help you find something today?* (Sometimes in that same chipper tone as the girls at Banana Republic.) *Looking for anything special today?* Everything is special today. If you wait long enough, a woman will come out from the back door behind a certain strip mall and make herself throw up on the grass.

I'm standing in the artificial-tree and holiday decorations department at the Macy's in Stonebriar Centre on a September afternoon. Christmas beat me here. Nat King Cole already sings "Chestnuts roasting on an open fire" to me, the only person around. After another slow lap through the mall, I walk outside to my car, and it's ninety degrees out, the ground baking under this Bethlehem redux.

Flocks and flocks of grackles fly in restless, swirling tendrils above us each evening, during the five o'clock traffic, settling briefly on the colossal concrete pillars of on-ramps and overpasses still under construction. They watch us in our cars. They shriek in unison. The nets on the trees in the Office Max parking lot are meant to discourage them, and haven't. Every few minutes they fly up again in a nervous burst and settle now on new ledges and new wires and new signs with new views.

This is where I'm searching for America's Christmas present. This is where I've disappeared. The star in the east would turn out to be a long line of jumbo jets in the lavender dusk, their bright landing lights aligned in near-perfect conjunction, on approach to D/FW. (*Radiant beams from Thy holy face. With the dawn of redeeming grace.*)

Often with Christmas, the easiest metaphor will do.

You look for it anyplace, and there it always is.

2

Town & Kountry

THE GUARD WAVES my car into Stonebriar Country Club Estates at a quarter to seven on the first Thursday morning of November. I wind through the neighborhood, along the golf course, turn right, turn left, and pull up to Tammie Parnell's big brick colonialesque house. Tammie is already waiting for me in the long driveway, next to her enormous, Coke-can-red GMC Yukon XL. "Are you ready, elf?" she asks in her high-decibel southern bark. She is wearing something fun — indigo jeans and fashion boots, a chocolate-brown turtleneck, and a suede vest trimmed in a foxy faux fur. "I love that you're on time," she hollers. "I love people who are on time and ready to go! Are you ready to be my elf? We are about to get moving!"

I am ready to be Tammie's elf.

Between now and Christmas, I will be Tammie Parnell's elf a lot.

When she is not trying to be the best, most involved mother and wife in all of Stonebriar Country Club Estates, Tammie has a business on the side: she does people's Christmas decorating for them, because they no longer want to do it themselves. She charges by the hour.

Tammie is taking a last look around her garage before we head off on a daylong shopping trip to the biggest antiques, arts, and crafts market in the state. It's a monthly event called "First Monday" (even though it always starts on a Thursday and lasts five days), held in a small town called Canton, about sixty-five miles east of Dallas.

Tammie's blond, bespectacled, preppy husband, Tad, is on breakfast and homework-in-the-backpacks duty for the couple's two children, Emily and Blake. He'll drop them off at school, a $20,000-a-year private Christian academy where Emily is in fourth grade and Blake is in second, and then continue on to his job as a vice president at the headquarters of Haggar, the clothing brand. (Tad oversees Haggar's accounts with J. C. Penney, which is headquartered nearby in Plano.) Tad does his best to stay clear of his wife's Christmas thing, a Yule frenzy that never seems to end. The Christmas retail season already controls a part of his work life; at home it has an even tighter grip.

The Parnells' immense three-car garage is occupied, year-round, by the effluvia of Tammie's obsession. There's an army of grinning Santa Clauses, elves, angels, Magi, and camels. Some are stuffed plushies, some are porcelain, and some are ceramic. There are forests' worth of plastic garlands kept in plastic bags, and stacked Rubbermaid tubs that are filled with ornaments, ribbon, tassels, and sashes. There are small artificial trees. There's a shiny lavender tree that is not for sale to any client, Tammie says. It's for Emily's room. ("Isn't it just the best?" Tammie says. "Hobby Lobby! Can you believe it?") Tammie has amassed a collection of matching decorative plates for serving cookies or snacks, and plates for displaying candles, and plates that are simply for show. There are small, medium, and large cones in bright green and gold, like dunce hats for misbehaving elves, that serve no other purpose than to look pretty on tables and sideboards or by the fireplace.

Dear Santa, reads a plaque in Tammie's laundry room, off the garage, *I can explain.* She loves the holiday so much she planned her and Tad's wedding around it in December 1994, dressing her bridesmaids in green velvet. She puts a red jingle collar around the neck of her black Labrador, Toby, even before the trees are lit in the home-and-garden department of Target. This could be the story of a woman who loves Christmas too much: she will work late into the night, every night for a month, to make other people's houses look perfect, and then, with Santa Claus's sleigh fast approaching, she will scramble to make her own family's Christmas appear more serene and wholesome, only to discover each year that she can't do it all.

"Movin', we've got to get movin'," Tammie says — when she's really excited, she starts dropping her g's, becomin' more twangy — backin' down the driveway in Big Red. (She named her SUV the day she picked it out.) "Beatin' that traffic, yes," Tammie says. "Everyone out of my way. We're makin' Christmas now."

Tammie is forty-four, slender, with deep brown eyes and thin lips. Unlike most of the supermom circle to which she belongs, she has not gone hot blond or under the knife or Botox needle for improvements; she has her hair done in hues of brown, in chin-length chops, with a few golden highlights. Think of Holly Hunter cast as a country-club homemaker.

Many days Tammie can be found at her kids' school, either volunteering or substitute teaching. At a moment's notice she can reorganize carpools with military precision on her Motorola Razr phone. She coaches Emily's volleyball team, the Smash Girls, with the energy level (if not the blue vocabulary or thrown folding chairs) of a courtside Bobby Knight. She rigorously inspects Blake's reading and math homework, because all these counselors and auditory-processing experts keep telling Tammie she's likely got one of modern exurbia's overstimulated boys, who are variously diagnosed with some level of inattention. If not on a baseball field or a golf course, Blake would prefer to have his eyes on his hand-held PlayStation whenever possible. There is no way, Tammie says, that she's letting her son fall behind. One teacher told her Blake wouldn't be able to do timed math exercises. "I told [her], 'You know what? He might not make As all the time or Bs, but we'll be happy with Cs if he's done his best,'" Tammie says, launching into it for me. "'So don't you ever, *ever* tell me what my child can't do.'"

Fiercely devoted, ever vigilant: Tammie shut down an entire water park one summer day when Emily and a friend disappeared for a half-hour. (The girls were hiding from Tammie, it turned out.) She worries from time to time about the world "out there," beyond the insularity of the kids' school, the other moms, Stonebriar Country Club, and Frisco itself, which Tammie knows, deep down, is not exactly the real world. Sometimes she volunteers for a group called Christian Hands in Action, which has sent her and other women on mission trips to Mexico for a few days to work in orphanages and dental clinics. These places way outside her world do a number on Tammie's heart.

When October comes, her Christmas craze ignites. The decorating business began three years earlier as a partnership between Tammie and a friend. They decided to call it Two Elves with a Twist and did just a few houses at first, finding their clients by word of mouth. "You know what was great about that partnership," Tammie says, "was that we were never friends before Two Elves with a Twist. We didn't really know each other, didn't have a lot in common, except we had boys the same age. We had a great time. She led a great life, I led a great life, but we just never—" Tammie pauses here, thinks about it. "She just never wanted to do more. I felt like at the end, that second year of the business, every time we were in a house, she kept looking at her watch and saying, 'Okay, we gotta be in and outta here in three hours.' And that's not what it's about. I want people to be happy." Tammie stayed in some of the houses long after her co-elf had left for the day, making sure everything was just so.

This year it's just the one elf. The official story, according to Tammie, is that the co-elf decided to go back to her full-time corporate job. (For all I know, the other elf's body is neatly bubble-wrapped in a Rubbermaid tub way, way in the back of Tammie's garage.) Tammie kept the name and logo — two mod cartoon women in Christmas-tree cocktail dresses trimming a tree — and doubled her solo effort, which tripled her business.

Tammie is already booked to do one or two houses per day, six days a week; her first appointment to install someone's Christmas comes up in the second week of November. Her last job can be squeezed in as late as December 11 or 12. (There would be something seriously wrong with a woman in these suburbs to not "have her Christmas up," as Tammie puts it, by then.) The jobs in her clients' homes take anywhere from a few hours — she'll charge $400 or so to do the living room only — to fourteen-hour marathon efforts to fully decorate several rooms in one of those 6,000-square-foot (or larger) Hummer houses, those red-brick or limestone-covered mini castles with Rapunzel-ready turrets in front and gabled porte-cochères leading to Garage Mahals. On these big jobs, Tammie recruits mercenary elves (her friends, usually, or other moms from the school), charging the client as much as $1,200.

Tammie is hoping to gross close to $30,000 by the time she hangs the last ornament in 2006. But she will have a capital outlay of several thousand dollars in materials and supplies — decorations she picks up at wholesale markets, warehouses, crafts fairs such as First Monday, or discount stores like Hobby Lobby and Michaels. Tammie resells her latest finds to her clients at a slight markup. She decorates a Christmas tree in her own dining room as early as September and invites other moms from school or church over for a merchandise party, where at a moment's notice she can show off several holiday centerpieces, arrangements, and mantel ideas. "Movin' the merch!" she shouts. "That's what I'm all about."

Tammie recognizes opportunity in another woman's meltdown moments. The ideal client is one who sees her house as an uninspired mess at Christmas. Tammie's clients will later tell me about how overwhelmed and helpless they felt until Tammie arrived in their lives. The sheer size of their dream houses gets the best of them. Before Tammie, they were still just putting old ornaments on trees, with multicolored strings of discount lights. They hadn't really given much thought to the dining room, besides some seasonal table runners from Pottery Barn or a berry-ball wreath candleholder from Crate & Barrel, or that same old crystal vase filled with red ornament balls or plastic pears, next to some random Nativity scene they'd picked up at a church bazaar in 1991. They hadn't figured out the mantel. They would get the garland wrapped halfway up their grand staircases before collapsing in frustration and heading out for margaritas at Gloria's with the girls.

On a first consultation, Tammie has a new client drag out all her cardboard boxes and Rubbermaid tubs of Christmases past from spare closets, extra bedrooms, garages, and walk-in attics. (These spaces are usually filled to bursting with the signs of full-blown affluenza: never-ridden bikes and hardly trod treadmills; abandoned lamps, vases, pots, gothic curtain rods; never-assembled organizer shelving systems; unheeded inspirational plaques; crafts projects attempted and deserted; half-collected collectible figurines; the husband's beer stein collection; boxes and boxes and boxes all marked "keepsakes.") Tammie will take a long look at the Christmas junk, zeroing in first on the key item: in what condition is the family's artificial tree? (She can tell right away if the owner had sprung for a quality tree at the outset or if it looked like a tree you'd pick up on sale at Wal-Mart. Tammie's rule on prelit Christmas trees is that anything less than one hundred lights per foot isn't worth assembling.)

Next, she wants to know what the client had been doing on her front door, porch area, and foyer. (A wreath? Of greenery or of decorative twigs? Ribbons?) What sort of Nativity scenes does she own? (This is also Tammie's way to ascertain, if she does not already know, the degree to which the house is, in her words, "Christ-centered.") What objects should go in the kitchen? (Decorating the kitchen for Christmas is huge in Tammie's world.) How to decorate the dining room table, sideboard, and chairs? What rooms upstairs will need decorating, that is, with auxiliary trees? How does the client feel about a tree in the master bedroom? (Tammie's favorite tree in her own house is the one she puts in her and Tad's bedroom.)

Many times the client will just shrug and tell her to do it all, signing off on a Tammie shopping spree. "Those are the ones who want to go off to work or wherever and come home and have it all done and looking fantastic," she says, "and they just want to write you the check." Which is fine with Tammie. It accommodates her idea that she is working magic.

Ever since America fell in love with a picture of Queen Victoria and her family decorating a German-style tree in their Windsor Castle home in 1848, Christmas has reminded women just how much everyone's holiday happiness (and unhappiness) falls to them. The Christmas lifestyle as most Americans know and celebrate it is only about a century and a half old, a straight line from Charles Dickens to Martha Stewart. The still-intact Victorian conventions of Christmas have Father worrying about money and security, and Mother saddled with making everything look and feel right — whether she has the holiday spirit or not. Despite advances in gender roles and the aid of modern conveniences, a happy Christmas still hinges on the woman's grasp of the evasive concept of *hearth,* a warm glow of peace and satisfaction. Christmas is a domestic crisis event, in which everything must be flawless, from greenery to stocking treatments to wrappings to elaborate meals. The anxiety begins in October, when the first catalog or magazine arrives featuring page after page of those perfect table settings, perfect cookies. The mailbox is soon crammed with images of amazing trees. Tammie's clients want nothing more than to live in a house that looks like a magazine; Tammie wants nothing more than to give it to them. So it begins with the tree.

Holiday trees of Tammie's world come straight out of the box, created in massive factories in China's Guangdong province, ready to fluff and plug in. To China's credit, these trees have never looked more real, and they are getting bigger and taller, as are the houses they eventually occupy. In and around Frisco and the adjacent megaburb of Plano, women (and the occasional husband) find new levels of exasperation each November in determining not just the style of a single Christmas tree for the family room, but also what sort of animal-themed or princess-themed tree to put in the children's rooms, and what sort of masculine, Dallas Cowboys or University of Texas Longhorns–themed tree to put in the media room, and what sort of elegant, ribbon-and-peacock-feathered tree to put in the formal living room. There are stories told — true or exaggerated — of gated-community women who have houses so large that they can keep fully decorated Christmas trees in spare walk-in closets and have the housekeeper just roll them out for display on the day after Thanksgiving. Well above Tammie's price point, there are big-name decorators who liberate trophy wives from the task of decking the halls, creating extravagantly themed home displays in time for party season. (My favorite high-stakes decorator in the Dallas area, if only for his inspiring name, is

Harold Hand.) Attuned to the requisite moments of picture-perfect togetherness, these decorators know to reserve one box of the client's "special" heirloom ornaments, left for her on an ottoman by the family room tree, so that she can have a few minutes of tree-trimming togetherness with her husband and kids.

The women Tammie works for tend to view Christmas decorating as an overrated chore but don't have $5,000 or more to get a Harold Hand. If their husbands are hiring Latino yard crews or off-duty firemen to put up the lights on the exterior (one local landscape company boasted 12,000 clients for its outdoor light-hanging business in 2006), what's so wrong with a little gal-on-gal consultation about the interior? There are dozens of other part-time decorators like Tammie, handing out their business cards at Bunco groups, Bible study groups, and soccer matches, who now earn $100 to $500 an hour (sometimes more) to deal with other women's ornaments, tangled light strings, manger babies, and plastic greenery.

For years Tammie has been clipping and archiving magazine photo spreads of fancy Christmas trees and garland arrangements. Through the entire George W. Bush administration, she harbored a secret fantasy of helping Laura Bush decorate the White House. From her special sheaf of Christmas ideas, Tammie produces a 2003 magazine picture of the First Lady standing by the tree in the East Room. "I have so many ideas for her." She sighs, certain she can do better than the entire fleet of hired help that puts together the White House's look each December. If only Mrs. Bush would call.

Tammie vows to teach me everything she knows about holiday décor, and I have vowed to listen. When Tammie says that strings of coral-like beads strung through plastic garlands or piled into golden plastic cups or red glass vases on dining tables and sideboards will be, in her words, "huge this year, you are starting to see these beads everywhere," I accept it without question.

"It's cute," I say.

She crinkles her nose at this disingenuous use of the most ubiquitous word in a modern woman's vocabulary. *Cute* to a woman in Texas is what *dude* is to young American men (or *shit* was to George Carlin, or *snow* is to Eskimo-Aleuts), with infinite ways to say the same word and inflect nuances in the meaning. A *cute* that is too clipped means, dismissively, *not that cute.* Tammie will teach me. My sense of cute is to be broken down and reassembled here, from scratch.

The first time I meet Tammie in person is at her church, at the St. Andrew United Methodist holiday bazaar in late October, where she has two tables of Christmas stuff for sale, actively persuading women to watch her drape coral beads over vases and candelabra. ("Does your mom make all this jewelry?" one customer coos to ten-year-old Emily, who replies, "No, it's all from China.")

The second time we meet, we sit on Tammie's back patio, near the slate-lined lap pool and spa Tammie and Tad just had installed, and we sip Chardonnay while Tammie flips through a fat binder of photos she keeps from her past jobs. In some houses, the client has pieces that she won't let Tammie dismiss. These items could be a giant, moth-eaten snowman from the 1970s, or a childhood sled, or something sentimental but unfortunately way too "Grandma," in Tammie's words. (An interesting and implied aesthetic demand by Tammie's clients is that their houses not resemble their mother's or grandmother's house too closely. The modern exurban woman fears the old-fashioned "country" look, even as she surrounds herself in expensive, more formal replications of it.) "I never force it," Tammie says. "You always have to work with it, because Christmas is really about those special things that we want around us, that are special only to us."

I ask Tammie if anybody has a real tree, or if she ever uses real greenery on the mantel.

Almost never, Tammie answers bluntly. None of it is real. She tells me to underline this one fact in my notes: "Fake is okay here."

It is faux-finished, plasticized, and derivative — but that's not the point. Tammie has Christmas figured out. It has less to do with true authenticity than a feeling of it. Women want big fake trees with lots of ribbon artfully wrapped in and out of the branches. These big fake trees should evoke certain themes: "old-fashioned" but "modern"; "Victorian" and "rustic" but somehow "original" and "new"; "homey" but not "amateur"; "rich" but not "cold."

"Absolutely," she says. "Fake is okay here. Diamond earrings. Christmas trees. If you want me to prove that fake is okay here, let's you and I go to the Stonebriar Country Club pool one day and check everyone out. You will see that fake is okay here."

Tammie finds it fascinating that I've never put up a Christmas tree in any of the apartments in which I've lived in the last two decades. (Tammie wonders, I can tell: *What kind of sad person has so little to do for his Christmas that he has to move here and watch me do mine?*) She instantly wants to help: "How big is your apartment?" she asks. (Eleven hundred square feet or so.) "Do you have a fireplace?" (No.) "Is your dining room separate from your living room?" (No.) She keeps prodding. "You will get so many ideas from me."

We continue paging through her look book. Tammie lingers on one photo: "It's one of my bigger houses. They live over in Preston Trail, do you know where Preston Trail is? Okay, it's an all-men's [golf] club, very kind of old-money. They're divorced. This is a huge house. He still has the house. She has a smaller house now, and I'm waiting to hear back from her" — waiting to hear back to see if the ex-wife wants Tammie to decorate her new, not-as-big consolation house. The husband, Tammie says, wound up with a younger woman from his office. She stares at the pictures from the 6,500-square-foot Preston Trail job, wistfully. "But, um, anyway, she had these *phenomenal,* phenomenal-looking Wise Men."

Absolutely and *phenomenal* are Tammie's two favorite words, used independently or together. The world reveals itself quite often to Tammie as absolutely phenomenal, except where it involves the specters of crime rates, high taxes, and all things Hollywood, including the song Emily was singing the other day: "Over and over again, 'I'll tell you my dirty little secret, dirty little secret.' I'm, like, 'dirty little secret'? No, that's it. We're not going to listen to that station in the car anymore," Tammie says. Things that aren't absolutely phenomenal include almost anything "negative" — cancer, autism, tackiness, and, finally, Tammie's sworn enemy, low self-esteem.

The absolutely phenomenal always prevails. The way a strand of prelit plastic garland adorned with stiffened tulle ribbon can look when tastefully arranged on a fireplace mantel in a formal living room is phenomenal. It could be *absolutely* phenomenal if a series of three honey-hued jeweled cones were also placed just so, leading the eye naturally to the ceramic manger scene on the dining room banquette, and, in the Tammie Parnell style, if you find just the right ornament balls to fill up the glass vases from Hobby Lobby on either end of the banquette. ("Half-off, *shhh,*" she says, conspiratorially, even when there is no one around. "I bought up all they had. The manager sees me coming, and he knows!") *Absolutely phenomenal* also applies to so many of the people in Tammieland — teachers, pastors, or self-made millionaires. Tammie knows a man who warded off cancer through a combination of prayer, attitude, walkathons, and, yes, well, radiation and chemotherapy. It could go without saying that he is absolutely phenomenal, but Tammie always says it: "Such a phenomenal man, absolooootely, and his wife is just phenomenal."

In the time I will know her, I rarely see Tammie do anything like rest, sit still, finish a thought, or listen intently without interrupting. She thinks it's important to know the details about everyone in her sphere: where the dad works, who the mom is (and whether the mom works; Tammie harbors a not-so-secret longing to return to corporate America, but her kids come first); where the parents went to college, where they go to church, what grade the kids are in at which school (private or public); which house they live in on what street in which subdivision; and, deeper still, what there is about the family that Tammie can brag about to others, and what there is to pity. She extends the story arc of triumph-tragedy-rebound to all. Couples have beautiful children, who overachieve, but they also have an impending divorce, or some nasty custody dispute that Tammie has a way of learning all about. People could have an absolutely phenomenal house with absolutely phenomenal furnishings, but the husband is caught having a gay affair or the wife has a cancer that goes in and out of remission. The daughter can get a college volleyball scholarship, but the son struggles with Asperger's syndrome. Tammie prays for all of them. It's like a marathon of *Bewitched* reruns where Tammie plays all the parts: She is a little bit Endora, sharply observant and a bit wicked. She is Mrs. Kravitz, nosy to a fault. But mostly she is Samantha Stephens, here to work domestic wonders. She has to know everyone's whole story, but only so their dirt can be made clean.

On our way to the First Monday market, Tammie merges onto the LBJ. The morning skies are now bright blue and the air is just crisp enough that she can smell Christmas about to happen. As she weaves into the fast lane, she gets out her phone to call her neighbor friend Kelli, to ask about another neighbor, Lorraine. They all live in Stonebriar. Lorraine has been fighting breast cancer for a few years. After everything she's been through, it looks like she is going to die soon. "The saddest story, you would not believe," Tammie says. "Phenomenal people. Two darling kids. The husband works at —"

Kelli's voice mail picks up.

"Hey, darlin'! It's Tammie!! Listen, what is going on with Lorraine's house? I left a message and I haven't heard back and I just really want to know how she's doing and where we're at. We need to figure out when we are going to bless her and get that house decorated for her. Call me! Bye."

This is Tammie with her Christian hand in action: she has promised Lorraine that she and Kelli will come put up her tree and garland and decorate her house for her, for free. "We told her not to lift a finger. She can tell us just how she wants everything to be. We are going to show up and make Christmas happen in that house. You just wait. It's my do-good this year."

Dallas recedes, and much flatness and sky pass by, and at a little before 9 A.M., Big Red exits into a mishmash of waffle houses, new motels, KFC, and Taco Bell. "This is it," Tammie says.

Canton, Texas: we are in the country, but we are also in what I call the *kountry,* an in-between place that thrives on repurposing old-fashioned America for new-fashioned America.

Kountry is the capital of middle-American décor. I am a native son of kountry, by way of the Oklahoma City suburbs, raised on gravy as a food group. Kountry exists everywhere, but to truly get to it, you have to be at least sixty or seventy miles from any chance of a Williams-Sonoma, Apple Store, or Barnes & Noble. Christmas may occupy a quarter of the U.S. retail economy, but in the true kountry, Christmas is a much bigger linchpin. In the kountry, for example, some people know that the Hallmark tree ornament catalog is released on the second Saturday of July, and they drive miles to the nearest Hallmark store that day to get one, the same way others in the kountry know the opening days for various hunting seasons.

Let go of "Less is more." Kountry preaches maximalism: more ornaments, more trees, more stockings, more decorative cookie plates, more holiday sweaters. Kountry dictates which lawn sculptures and bric-a-brac will catch on next. Kountry continually reinvents the sachet, the frilly coat hanger, the hand-painted pillowcase, and the plaque with the pithy Bible passage. It lets you know whether roosters are more popular than ducks this year, or if it is to be Holstein cows again. The kountry economy was the first to start selling T-shirts that say *What happens at Meemaw's house stays at Meemaw's house.*

Kountry keeps some Americans firm in the belief that we are culturally rooted in the rustic. It made "southern living" possible anyplace. It gave us everything from Grandma Moses to Paula Deen. It sends designs of antique-style, kountry objects to Asia to be mass-produced on the cheap, and then resells them to the suburban citizens whose Meemaws used to own one just like them. The kountry economy will never run out of ideas for what to do with broken farm implements, old soda bottles, pieces of barbed wire, and hard-sided suitcases. In fact, ever since the marketplace ran short on the actual detritus from nineteenth- and twentieth-century small-town American life, it began importing from abroad all manner of kountry reproductions, including old Santa Claus Coca-Cola advertisements, barrels, birdhouses, windmills, and pieces of wood treated and distressed to look as if they came off old barns. Kountry is both supported and despised by the prevailing aesthetics in contemporary American interior design; yet even doctrinaire minimalists in the city can be frequently spotted at events like First Monday, sneaking around in Chanel sunglasses, on the off chance that something authentic (or cool) will accidentally present itself.

First Monday occupies several hundred acres of former pastureland just off Interstate 20 and monthly attendance numbers in the tens of thousands. First Monday has its own newspaper and governance; sprawling RV parking lots accommodate the vendors and customers who travel here and make kamp all five days. The market area, with gravel underfoot and a 4-H hint of livestock stable smells, features pavilion shelters the length of several city blocks, where dealers rent space. The entrepreneurs here do business under names such as Razzl Dazzl, Appli-Kaye's, Good Old Stuff, Santa's Ranch, Thinkin' Pink, Aged with Grace, Debbie's Closet, Quiltin' Cousins, Mary's Makins, and Nuthin' But Ribbon. Further in, there's a flea market that appears to stretch into the next county, where the First Monday shopper can find marked-down knockoff handbags, Lacoste polo shirts, sunglasses, hunting knives, African drums, and Native American rugs. First Monday covers so much area, and the citizens of kountry are so often disinclined (or sometimes unable) to walk very far, that there is a booming business in electric scooter rentals. One heavyset man glides slowly by on his rental scooter while pulling a small flatbed trailer, upon which his even heavier wife is perched in a plastic patio chair, giving him directions to the booth she wants to stop at next.

Tammie figures it will take us about four hours to work

through all of First Monday's Christmas offerings, including a quick lunch stop at her favorite vendor in the food court for a baked potato and a slice of coconut pie. "I like to do the big buildings first. We're looking for a little bit of everything," she says: small matching trees, lots of bejeweled cones (as in dunce hats, not pinecones, though the right pinecone can come in handy), gingerbread men of any shape and size. Also her heart melts when she sees little elves (plush or figurines), and she must have them. Last trip, Tammie struck gold at the deeply discounted booth of a popular home accessories outlet store called Paul Michael and came away with plates, small decorative boxes, vases, and small, medium, and large versions of the dunce-hat cones. Tammie also needs ribbon — she prefers floral ribbon with wire edges, pieces of which factor into many of her Christmas creations, whether on a tree, banister, or mantel.

Kountry welcomes Tammie and her Capital One credit card. Right away she finds a stuffed elf that, if he could stand, would be almost four feet tall. Button eyes, yarn smile, thoroughly creepy. Tammie envisions him sitting in about five different living rooms on her client list. "Oh, he's a *bad boy*," she says, lifting the elf up. "Oh, I can use him. What do you think?"

"I think he's . . . fun," I lie.

"I can tell — you don't like it. Be honest with me. You're my other set of eyes today, friend. You have to tell me."

We move on, the stuffed elf with us. (Pretty soon, Tammie has to buy two wheeled handcarts to carry it all.)

One of Tammie's most important stops is at a large booth run by a Korean family that sells dozens of varieties of plastic garlands, greenery, twigs, branches, fake fruit, holly, berries, and numerous combinations of all of the above. The American kountry Christmas involves a constant, evolving array of alien foliage, things not ever seen in nature. For example, frosted cherries and oranges grow on pine branches, along with bright orange berries of no determinate fruit. Twigs are whipped up, as if by some banshee wind, into cones and spheres. Bare plastic branches bloom with bright red and lavender berries. Glazed, sugar-encrusted plastic pears grow on plastic ivy, on the same vine that blooms with glazed apricots and red apples. Antlers are growing on branches and vines. Tammie loads up on a couple of hundred dollars' worth of Things Not in Nature.

Every object Tammie picks up — "What do you think? Is it too-too?" — she turns over and examines for any sign of a label. Many Christmas ornaments and accessories she sees bear the label of a company called Raz.

"Raz, Raz — I always see this name, *Raz*. They really have the market cornered," she says.

I later learn that Raz is a somewhat clandestine operation, based just west of Dallas in Arlington, Texas. A representative at Raz's headquarters explained to me that the owners "like to remain anonymous," and she would give no hint of annual sales figures. The company specializes in decorations — for all holidays — which are designed by a few staffers in the United States, who then consult on prototypes before each item goes into mass production in China or in other countries. The ideas for a Christmas knickknack can come from anywhere, my source explained — even from telephone operators who work at Raz's catalog switchboard. Raz designers comb church bazaars and antique markets in the South, the plains states, and New England, seeking the latest farmy-charmy innovations while also getting a read on what's been done to death and no longer sells the way it used to sell. A Raz-made object meant for Christmas 2006 would have been conceived and designed a year or two earlier, endured the prototype phase, and then gone into production, after which it then made the rounds at various wholesaler market shows in China and beyond, winding up in year-round Christmas stores and bazaars spread o'er the hills and off-ramps of our kountry.

By noon Tammie has purchased more than she and I can cart. We hike back across the highway to Big Red, unload our booty, and return for more. After one last pass, two hours later, we stagger back and cram glassware, wooden bowls, a couple of dozen mini tinsel trees, elves, cones, ribbon, and a forest of Things Not in Nature into the back half of the SUV. Tammie has spent close to two grand.

I'm surprised there's room for Blake and Emily in the back seat of Big Red when we pick them up, but they wedge themselves between garlands and mini trees, wearily, as if it happens all the time. On the ride back to Stonebriar Country Club Estates, I ask them what they want for Christmas.

"A dirt bike," Blake says, instantly.

"Um, a laptop, I think?" Emily says.

"We'll see, won't we?" Tammie says, making eye contact with both of them in the rearview mirror, merging back onto the Dallas North Tollway at a rush-hour crawl.

3

It's Bazaar

TAMMIE HOSTS AN open-house party a few days later, November 7, spreading Christmas cheer all over her 4,000-square-foot home, displaying stuff she'd be only too happy to sell: trees in her dining room and living room, vases and arrangements and wreath treatments in the dining room and foyer, holiday knickknacks in the kitchen, lots of smiling Santa Clauses in the family room, and cones and other gewgaws along both fireplaces. Tammie says all this is just for today's show; she will completely redo everything for her family's real Christmas.

She has also invited two other enterprising businesswomen. One co-owns a company called Two Funny Girls, selling holiday- and sports-themed Styrofoam party cups—"Gobble 'til you wobble" reads one with a turkey on it; "Weekend Waterford" reads another. The other self-starter is stirring soup on the stove and baking muffins in Tammie's kitchen—part of her job as a sales consultant for an outfit called Homemade Gourmet, which sells assorted easy-to-prepare meals and powdered cake, cookie, and muffin mixes. The Homemade Gourmet lady goes on at length about her product line, and about the woman who founded the company on Christian principles. "She doesn't want to just sell food," the soup stirrer informs me. "She wants to change America. She wants to change values for the better."

Christmas in the suburbs brings forth this curiously heightened sense of the solo-entrepreneurial, beyond the usual world of irksome neighbors all trying desperately to sell one another Pampered Chef kitchen supplies and Mary Kay cosmetics. It suddenly feels like everyone in Frisco is selling anything and everything with some Santa on it. Some have day jobs and sell stuff on the side. Others do it full-time, stockpiling their wares all year for the October church bazaar circuit. For some it's a hobby gone haywire, often with Jesus as the unseen business partner.

On every Saturday in October, I go to a different church holiday bazaar, where someone is always selling faith-based scrapbooking materials licensed by the Once Upon a Family line. They're selling nutritious smoothie mixes boxed in healthful, family-centered packaging; they have licenses to peddle Christ-centered body lotions and face creams. They make jewelry in their spare time — cross-shaped belt buckles, cuff bracelets with scripture passages. They sell baby clothes bought wholesale from suppliers, under business names like Angel Darlin'. They sew Santa onto almost anything, especially denim vests they also buy wholesale, BeDazzling them for resale at a higher price. In garage workshops, men will cut and paint plywood into manger animal shapes and approximations of Disney and *Peanuts* holiday characters to sell to you to put on your lawn. Spare-time artists paint detailed watercolor depictions of family trees, with all your relatives added to the branches (and the ones you don't like left off), the perfect gift. One woman at a table at a church bazaar is selling tiny clear bags of small marshmallows, onto which she's stapled a construction paper snowman and the following greeting: "You've been naughty so here's the scoop. All you're getting this Christmas is snowman poop!" She wants $2 apiece for them. From a few feet away, they look like little baggies of crack, as seen on *Cops*. In a way, it's all crack.

But handmade, right? Local, correct? After listening to a woman at another church bazaar talk about how hard her daughter worked painting these lovely Santa figurines, I ask why this one has a "Made in China" sticker on his backside. "Well, all but *those*," she corrects herself. "Actually, that's not our stuff there, I'm selling that for a friend. She does those."

She looked at Pa sitting on the bench by the hearth, the firelight gleaming on his brown hair and beard and glistening on the honey-brown fiddle. She looked at Ma, gently rocking and knitting. She thought to herself: "This is now." She was glad that the cosy house, and Pa and Ma and the fire-light and the music, were now. They could not be forgotten, she thought, because this is now. It can never be a long time ago.

— LAURA INGALLS WILDER, *Little House in the Big Woods*

At Tammie's urging, I prowl around a weekend-long event called 'Neath the Wreath at the Civic Center in Plano. It's the Junior League chapter's holiday bazaar, held every November, featuring 108 merchants, each carefully selected by an intrepid committee intent on laying waste to any other church or school bazaar. The event attracts 10,000 customers from Friday to Sunday. The Junior Leaguers have transformed the drab convention hall into a glossy winter wonderland of high-kountry expression, aimed straight at the niche market of women in their thirties and forties who've all taken to calling themselves "divas" (or "princesses," unless they've passed that title on to their daughters). Here is a ready-made market for "Queen Bunco Bitch" shirts, or cake servers shaped like leopard-spotted stiletto heels, or mammogram-advocacy key chains that say "Gram Your Mams." At an opening party — "Ladies' Night Out" — there are several hundred women in their sparkly, skimpy peasant tops and empire-waist low-cut blouses and tight pants, with their blond hair teased up or straightened to a sheen, squealing at one another about how cute they look, how cute this is, how cute that is, and did you see these — these diapers that say "Drama Queen"?

CUTE!

You hear it all over the room: That's soooooooooo cute. *Cyyyuuuuuute.*

I'm not the only man here, but almost. I slam a few complimentary shots at the Wine-a-Rita booth ("It's like a Merlot Slurpee, ain't it?" the gal behind the booth enthuses), and then I push through the gushing rabble of cute. The ladies coo over party dresses, skirts, shawls. ("As seen on Oprah — STRETCH BELTS, one size fits most!") Here are tabletop Christmas trees made from rusted metal (*cute* rust, not too orange, not *too* rusted). Here's a plaque that says "On Dasher, on Dancer, on Master, on Visa." "Here!" shouts a woman working at a booth filled with nothing but ready-made snacks. "Try this divinity!" (And this fudge, and these cheese dips!) Every booth is run by a self-made business with a darling name: Two Sisters and a Cat. Sacred Pause. The Pajama Princess. Experiences! Brushes & Bows ("a girly-girl boutique featuring head-to-toe apparel for the littlest princess and her tween sister!"). There is the Bead Lady, Annette's Touch of Class, and Mommy Made It. There is 2 Cute, and there is 4 Home.

'Neath the Wreath's most popular fundraiser — tickets have been sold out for weeks — is the American Girl luncheon and fashion show on Sunday afternoon for daughters and their parents and grandparents. Aged toddler to about ten, the girls arrive for the luncheon with their hair curled and done up in big bows, dressed in sweet velveteen and plaid church dresses or other frills — much of it straight from the American Girl catalog. The American Girl doll phenomenon is its own separate universe of cute, appealing to parents who rave about its wholesome and educational appeal. These parents brandish American Girl as a silver cross against the nefarious, limo-riding, tarted-up Bratz dolls. Mother-daughter trips to American Girl stores in New York, Chicago, or Los Angeles are seen as worthy and special pilgrimages. (A Dallas store opened in 2007.)

Each American Girl doll represents a different era in U.S. history — the colonists, the pioneers, Native Americans, the big wars. Each comes with her own narrative of an adversity overcome by ingenuity and heart: In the Great Depression, an American Girl could never throw anything away. In the 1800s, an American Girl wore her nightgown all day long, under her clothes, because she had no underwear. One American Girl managed to escape slavery via the Underground Railroad — and can you believe it? She owned only *one* outfit. "It's the sparkle, spirit, and style of American Girls, yesterday and today!" intones a recorded narration, as the lights go down.

A Junior League member and a teenage beauty pageant winner emcee the fashion show. While each young model, carrying a doll, takes her little turn on the catwalk, we learn her American Girl backstory. Here's Josefina, who lived on a ranch in northern New Mexico in the 1820s. She had to sew her own clothes.

"Who here knows how to sew their own clothes?" the emcee asks. "Raise your hands."

In a room of several hundred families, nobody raises a hand.

"Moms? Anyone here ever sew? Anyone have a sewing machine?"

No hands.

"Well, then, you can just imagine how hard life was."

This is a guiding-light principle of life in the kountry Christmas: reminding the modern consumer how special, how difficult, the past was. (Popcorn strung on a sad conifer that Papa cut from a hillside, down in the holler. An orange left by Santy Claus in your stockin'. *This is now. It can never be a long time ago.*)

After having their pictures taken while clutching their American Girl dolls in front of 'Neath the Wreath's fourteen-foot-tall Christmas tree, each of these girls will leave the Plano Civic Center with Daddy and slip easily forward into present time, whipping out a little pink phone as she climbs into the family minivan. Mommy and Mommy's friends stay to troll the booths at 'Neath the Wreath, woozy on Wine-a-Ritas. They search for charming new ways to hang stockings. They seek out brand-new "vintage" ornaments. They sample ready-to-bake "homemade" cookie mixes, cider packets, and potpourri concoctions, the very things that will soon fill their houses with some lost aroma of the real.

4

There Glows the Neighborhood

I F I COULD TRADE places with anyone in the 75035 or 75034 Zip Code, I would be a child again, just old enough to keenly observe the world around me and have it work to my advantage, but just young enough to make a convincing case that I still believed in Santa Claus and Jesus Christ as separate, if conflicting, omnipresent forces. My sneakers would have blinking red lights in them until I was six or so, on the infinitesimal chance that I'd dart into traffic at night. (Traffic? Night?) After a ketchupy, nuggety dinner at which any food item that offends me has been altered or removed from my sight, I would take the most amazing bubble baths in whirlpool tubs, after which I would be tucked in by 8:30, to have a story read to me with a sense of dramatic wonder by an incredibly handsome man who kisses my forehead and cheeks.

By day I would travel like a celebrity, chauffeured to a variety of carefully planned fun. At birthday parties, my friends and I would get manicures at kid spas and makeovers at Club Libby Lu, or leap into the moonbounce that always seems to have just been inflated for our pleasure. Once I'd outgrown my blinking-shoes phase, I would wear Heelys sneakers with built-in wheels, so I could roller-skate anyplace there are floors, rather than walk. Everywhere we Frisco children roll, there would always be a hundred children just like us, just as special, often with the same names.

Photographs and videos of me, carefully edited to present me at my most darling, go up on limited-access Flickr and Vimeo accounts, mostly for my adoring groupies (grandparents) who send constant fan mail and gifts and have even bought and sold real estate in order to live a couple of hundred miles closer to me. I get fingerprinted one day at the mall, which will aid my rescue when a kidnapper strikes, as kidnappers are surely about to do, seeing as I am so desirable. At school, specialists stand by to diagnose any learning difficulty or soothe any slight to my ego.

I'd have the latest in everything, except freedom to explore. My school and activity schedules trump all else in my parents' lives. Indeed, the entire city, even the shopping mall, is really a tribute to my world, and any business that does not cater to families will soon close. By late October, we children of Frisco start drafting long letters to Santa, sometimes so specific that we copy down inventory SKU numbers from retailers' websites, or, better still, just text him the URL links.

People without children are viewed with suspicion by other grownups. When a house is purchased by a couple who are in their thirties but don't have children, one of the street moms will likely express real regret about this to the other neighbors, as if the property has become a toxic spot on the cul-de-sac.

But there is one exception to these gross overgeneralizations. There is a house in Frisco occupied by a young husband and wife who don't have kids and don't ever plan to. Come Christmas, among adults and especially among children, that house is the most popular in town.

* * *

Everyone keeps asking if I have seen the Trykoski lights yet. If this is a book about the holidays and the setting is Frisco, Texas, then people insist — *insist* — I drive by some night and see the Trykoski house. "But it takes a long time," warns one woman. "We sat in traffic for half an hour."

I have seen the Trykoski house. I saw it last night and I will see it tomorrow night, and almost every night between now and the end of December. Then, in January, when all the lights are put away and gone, I will still be hanging out with Jeff Trykoski and his wife, Bridgette.

I've been with them since we first met in October, and since the big install on a Saturday in early November. That's where I am now, kneeling on their front lawn, snapping together white plastic snowflake lights while the Aggies football game is on the radio, which is connected to the day's first extension cord. The beer is in the ice chest.

I am helping (watching mostly) while Jeff and Bridgette and a dozen of their relatives and friends put up the 50,000 or so Christmas lights on the Trykoskis' one-story, three-bedroom, red-brick house — a house indistinguishable from the hundreds of houses surrounding it, until Christmas. Jeff's two brothers are here this afternoon, and so are Bridgette's mom and dad, and her brother, sister-in-law, and eighteen-month-old nephew, along with some of Jeff's coworkers, and some of his and Bridgette's friends from college days.

Over time, Jeff will show me more than just the difference between C-6 and C-9 bulbs, or how the whole thing is wired through circuit boards that are mounted to the south end of the house and how the wires are then run through the attic and controlled by the Dell hard drive in Jeff's study off the entry hall. Jeff will explain how the song that goes with the light show ("Wizards in Winter" by the Trans-Siberian Orchestra) is programmed second by second to play along with the blinking;

how the software controls the on/off switch of every connected string. I'll learn how the song goes out to passing cars via a limited-range FM transmitter.

I've been to the Trykoski house so many times now that I'm not even sure why I still ring the doorbell before I enter, except for good manners. I go Christmas shopping with them. I do Jell-O shots at Bridgette's birthday party. Once you're inside, you forget the mad light show going on outside for the parade of slow-moving traffic, every night from six to ten, from December 1 to December 29.

Life does not really change when the lights are on, Bridgette observes. Some nights get so warm still she wishes she could turn on the A/C, but she can't until 10 P.M. Same goes for the dishwasher, by orders of Jeff, or else the main breaker might flip and everything will go out.

One time, the doorbell rang and a woman stood there. It was late — a half-hour after the computer had automatically switched the lights off for the night. She had her kids in her car and asked the Trykoskis if they'd turn it all back on, just for a little bit. *Please,* the woman said, *we've just had the worst day. It would really help.*

Jeff turned them on.

Bridgette recalls how the family sat out there in their car, as the house and lawn glowed in a blinking tizzy, with the radio tuned to "Wizards in Winter." When you sit in front of the Trykoski house, this is what you see:

It is a show of dancing light. It begins as a pouncing, pulsing onslaught of light and music, growing denser and then pulling back, and then pouncing and bouncing again. From one edge of the property line to the other, the lawn is gridded by evenly spaced strings of multicolored lights blinking up, down, back, and forth, like a Travoltian disco floor. The shrubbery and live-oak tree are also covered in lights. A thirty-foot pole (Jeff had a permanent concrete foundation poured to support it) is circled by strings of lights that make a Christmas tree which also blinks synchronously — artfully, in sections — to the music, echoed by the off-and-on motion of two dozen snowflake lights on the roof. The roof edges are strung in larger white C-9 bulbs, as are the entryway and front windows. A big star sits atop the chimney. To all this Jeff has added a red-, white-, and blue-lit ribbon (a show of patriotism) and planted the flag of his and Bridgette's alma mater, Texas A&M. (It comes as no surprise to anyone with a working knowledge of Texas culture that something so odd yet ingenious, so difficult to execute, and so essentially geeked-out is brought to you by an Aggie.) Finally, the front of the lawn is lined in short plastic Christmas trees with mini lights. Five reindeer mechanically graze on the light-gridded lawn. All of this blinks with such staccato frenzy — note for note with the song — that it looks as if the switches are controlled by a maniac child. At first take, it is not a gentle experience, but then, as the eyes and ears follow along, it becomes something sublime. It becomes something you want to watch over and over.

Bridgette never went out there and asked if the family was okay, and what it was that brought them there, so late, or why the mother seemed so dejected. Bridgette can imagine. Bridgette DVRs *Oprah* every day. It could be about a child. It could be about divorce. It could be that someone's sick or the dog just died. It could be anything, for anyone, in any of the cars that idle around the Trykoskis' cul-de-sac in December. They come by to feel happy, but why else do they come? Bridgette wonders if there isn't something deeper to it. Telling the story about the woman who rang the doorbell that night always makes her eyes brim with tears. Is it possible, Bridgette wonders, that there's some bottomless need here that people have? For Christmas lights?

Inside the house — while Bridgette and I watch *Dancing with the Stars* and her brother-in-law, Greg Trykoski, drinks a beer in the recliner and picks fights with Toby, her temperamental cat, and Jeff works on his two computer monitors in the study — Bridgette sometimes forgets the lights are on, facing out to the world.

Seen only from your car, the inside of the Trykoski house is a mystery.

What are the people in this house saying?

Look at us, look at us, look at us?

The electric Christmas light goes back almost to the earliest mass production of light bulbs. General Electric acquired the patent to sell Christmas lights in 1890 from Edison Electric after several prototypes were tested, including one tree lit by an employee of Thomas Edison in 1882. By then the late-December house fire was an epidemic occurrence; though prone to sparking danger, electric lights somewhat lessened the anxiety that came with balancing burning candles on the dry branches of indoor Christmas trees. Grover Cleveland's White House was among the first to switch to electric tree lights in 1895. Affordable outdoor lights became available to homeowners by the 1920s, but the electric bills for a display could be prohibitively expensive.

In the Boyle Heights neighborhood of Los Angeles in 1936, George Skinner put 7,000 lights (and four 500-watt floodlights) on the outside of the bungalow he owned with his father. Skinner persuaded the power company to donate 80,000 watts of power for his "Christmas House." He later added music with loudspeakers. He even had eight railroad cars of real snow from Utah delivered, which was spread on his front lawn. Over three seasons, he claimed to have attracted 100,000 drive-by visitors. The whole point — the only point, his daughter Georja wrote in *The Christmas House*, a family memoir — had been to make people feel better. (George, a polio survivor, felt the world needed cheering up.)

More people put lights on their houses starting in the 1950s, as energy efficiency improved and they became affordable enough to turn on several nights in a row. Eventually the act took on a mythic reputation as one of the dreariest and most pratfall-prone chores of the season: unraveling the lights, testing the lights, getting out the ladder, hanging the lights just so, trying to figure out why the lights don't work, and, after all that and three or four weeks' time, getting out the ladder and taking them down again.

Hanging the lights is also the one part of America's domestic-centered, women-do-it-all Christmas that is commonly regarded as the man's realm. Pop lore both celebrates and mocks the kind of man who not only relishes putting lights on his house but also tries to outdo his neighbors. Chevy Chase's hapless Clark Griswold, in 1989's *National Lampoon's Christmas Vacation*, stands for all men who have gone too far with their house displays. ("Is your house on fire, Clark?" "No, Aunt Bethany, those are Christmas lights.") From this came the sitcom predictability of it, the jokes about the overambitious doofus on his roof who knocks the entire town off the power grid, or singes the dog, or zaps himself into a stuntman's arc, landing three or four lawns over.

Jeff will grant you that none of these comedies ever gets it right, but not because of the Griswold stereotype. The real problem is they never show you the schematics, or where everything is plugged in. They never tell you what the amp draw is. Hollywood would have us believe that Danny DeVito, in the drecky 2006 comedy *Deck the Halls*, can visit some quaint, village-style hardware shop and emerge with enough boxes of lights in his arms to completely cover a two-story house and yard, in an afternoon, by himself, without ever once drawing and redrawing his routing plans, without carefully numbering his extension cords. And really, who would believe that?

As the Christmas of 2006 approaches, Jeff and Bridgette are both about to turn thirty-one. They met when they were both working in Texas A&M's campus bookstore as undergrads in the late 1990s. Bridgette was a study hard/party hard biz major. Jeff was a computer guy, irrepressibly nerdy, an Eagle Scout who doesn't drink.

Bridgette has fair skin, bright bluish green eyes, and blond hair razor-cut into short, chin-length fringes. Jeff has brownish green eyes and an easy grin, and is losing his hair. Bridgette is a clotheshorse who keeps a mental list of pros and cons about the area's malls — Stonebriar Centre in Frisco (mostly cons) versus the Shops at Willow Bend in Plano (mostly pros) — and a list of reasons why none of the jeans or tops in her closet are right for the dinner or happy hour she's planning on a given weekend. Jeff wears the same thing every day — clothes Bridgette buys for him at Christmas or when she sees them on sale: "Khaki pants, brown shoes, brown belt that goes with the brown shoes, shirt from Dillard's that goes with brown shoes, brown belt, khakis," she says, about his Monday-to-Friday wardrobe. (Weekends: stonewashed blue jeans and Aggie sweatshirts.)

After graduation, Jeff took a job as a systems engineer at the Plano offices of Frito-Lay and PepsiCo. His business card reads *Jeff Trykoski, Lead Specialist: Metadata.*

"Metadata is data about data," Jeff helpfully explains.

Bridgette shrugs.

She has a job at Dow Jones's Irving, Texas, office, working in the advertising operations of the *Wall Street Journal.* It takes

Bridgette an hour to drive to work and an hour to drive home in her old Mercury Cougar; some days she says she screams at other drivers the whole way. It takes Jeff a few minutes to drive to work in his new Honda Pilot SUV, a trip he has timed down to the exact second, based on the patterns of two traffic lights. Jeff is always counting, figuring, adding, subtracting, and dividing. "When we're driving, he counts the mile markers and stuff," Bridgette says. "He counts in his head, compares it to the mileage [that the car is getting] per gallon. He doesn't even notice that the radio isn't on." She married a man who is his own global positioning system, who always knows where he is and where everything else is. "He is constantly calculating stuff in his brain. Always. It's amazing he sleeps so well. He just passes out. His brain can somehow shut off."

In Jeff's world, you would always, always label the top and bottom end of a new extension cord and assign it a number with a label maker, because God help you if you lose track of where an extension cord is going in a 50,000-bulb house display. In Jeff's world, you would have long ago ceased to write on all your Rubbermaid tubs with a Sharpie, because the *smarter,* more metadata thing to do is print out standardized, clearly marked inventory sheets that are *taped to the tub,* on which you can check off the contents of each: strings of Christmas lights with C-9 bulbs versus strings of mini clears, mini colored, multi-minis, and so on.

Logic, order, *meep-meep-meep.* Jeff is frequently beaming in and out of his world and ours. Bridgette will be talking about something, and Jeff will interrupt with some thoughts about how many mini lights it would take to cover all the lawns on their cul-de-sac and make it into one monster grid. Or he'll suddenly announce exactly how many times his lights blink on and off in the four hours that they're on each night.

"Jeff!" Bridgette says in her "rough" voice (she does sweet and she does mean, sometimes in the same sentence). "That has nothing to do with what we're talking about."

It takes me a while to read the affection in Bridgette's frequent snapping at Jeff. She is constantly yanking him out of some 2-D schematic and into a world where life doesn't line up the way you designed it.

The Trykoskis' address is 4015 Bryson Drive, the second right turn off Hillcrest Road into a mazy, moderately priced neighborhood called Hillcrest Estates, where all the streets are named after the children of the land developers and homebuilders who made it. Besides Bryson, there are Sara Drive, Max Drive, Hunter Run, Becca Circle, Sean Drive, Mallory Court, and so on. By outer-Dallas standards, 4015 Bryson Drive is already too small, at three bedrooms and 2,470 square feet (if you include the garage), and already ancient to the market, having been built in 1995. But it has everything Jeff and Bridgette Trykoski wanted when they were newlywed shoppers in 2001. They weren't looking for a baby's room or a backyard, a house where they could raise a family. (Jeff and Bridgette are the only exurbanites I've so far met north of the President George Bush Turnpike who have a plainly stated, unashamed desire to never have children.) They wanted a big *front* yard.

As long as Jeff can remember, he wanted his house to be *that* house at Christmastime, the house the neighbors talk about and tell jokes about how it can be seen from outer space; the house with hundreds of people waiting to see it every night, in traffic backed up a half-mile away. When the TV stations come each year to do wacky-dude-with-the-crazy-Christmas-house stories (as they always do), Jeff will, in his standard monotone sound bite, tell them how good it makes him feel to make other people feel happy, especially kids. "It's my way of giving something to the community," he'll say, or "It's what Christmas is all about."

No reporter ever contradicts that, because we're all supposed to know what Christmas is all about, and the sentiment should suffice, even as I watch Jeff give television interviews or read articles about him in the neighborhood sections of local newspapers and feel it barely ever gets at the *why*.

The neighbors all got used to it several Christmases ago. The kids who live across the street will sometimes set up a booth and sell hot chocolate to the people in the waiting cars. But one of the neighbors can't stand it, because of the traffic, and these particular neighbors no longer speak to the Trykoskis. Jeff is fond of pointing out that there's nothing in the Hillcrest Estates Homeowners' Association rules about Christmas lights because there is no Hillcrest Estates Homeowners' Association, which is precisely why he sought a house in this neighborhood. "If we had an HOA, I would have had big problems by now, because of the traffic thing," he says.

The Saturday in early November 2006 when we install the Trykoski Christmas lights happens to be Jeff and Bridgette's sixth wedding anniversary. "Just how I love spending it," Bridgette says, with another of her pretend sneers, meaning she does sort of love it. She's into the lights, too. She walks across the cul-de-sac with a bottle of Coors, to admire the day's work in the fading autumn light. Her husband is on the roof, and so is her father, and so is her brother, and so is one of her brothers-in-law. These are the men in her life, and this is what they do. She shouts out which snowflakes look crooked. "More to the left." "No, up a little." Finally it looks right. "That's good," she says.

Jeff is the oldest of Jack and Marie Trykoski's three sons. He was born ten weeks prematurely in January 1976. He weighed a little over two pounds, and it was fifty-three days before they could bring him home.

"The day he was born, they told me he was dead," Jack remembers.

"I never made a big deal of it," Marie says.

Jack works in the oil business, which has taken him all over the world — Europe, the Middle East, Asia. The Trykoskis lived in London, then in New Jersey, and then for most of Jeff's childhood in the piney suburbs of Houston, and now an hour north of Houston in the country of kountry, three and a half hours from Jeff's house. They had another boy, Greg, and another boy, Doug, all spaced two years apart. Jeff was the boy who looked forward to the arrival of the telephone repairman and would follow him around to give wiring advice. A lemonade stand could never just be a lemonade stand, Marie says: "It had to turn into this major corporation."

The phones were always in pieces; the wheelbarrow was disassembled for parts. These were the boys who taught themselves to program computers, who'd asked Santa Claus for a modem before any of the neighbors had one. "Jeff would lie in bed at night and read computer manuals," Jack says. "Who reads the damn manual?" Marie saw that the boys had rearranged the furniture in their bedrooms and discovered they were trying to hide the holes in the wall where they'd installed their own phone and modem lines. Jack says he'd pull up in the driveway and the two older ones would have the little one dangling on a rope, installing Christmas lights on the house. Or he'd get home and the boys would somehow have lights wrapped forty feet high around the pine trees. "Why'd you let them do that?" he'd come in and ask Marie, and she'd throw up her hands: "They said they wanted to put up the Christmas lights."

"It went from 'Can I help you put up the lights?' to this whole production," Jack says.

Jeff came down the chimney once.

"Don't do that when I'm not home," Marie told him.

(It *wasn't* the chimney, Jeff says, later correcting his mother's flair for Christmas metaphor. It was a tiny space between the drywall and the fireplace; he was installing a phone line.)

Having watched her sons become men, Marie would now like a grandchild. This desire grows more vocal each year. (Jeff and Bridgette have been clear with everyone that parenthood is not for them.) Marie also wants all her boys home for Christmas, the way mothers do.

Greg, who is twenty-eight and lives near Jeff in Plano, and his girlfriend, Christine Meeuwsen, come home for Christmas.

Doug, who is twenty-six, and his fiancée, Traci Wright, who live in Houston, always come home for Christmas.

But Jeff and Bridgette do not come home for Christmas anymore.

"We've explained it to my parents over and over," Jeff says. "We have to be at our house for Christmas, because of the lights." There's more than $10,000 worth of sixteen-channel control boards on the side of the house, and the house lures the most traffic from December 23 through 26. It's the Christmas that Jeff and Bridgette have made for themselves, and they are happy to share it, but they are staying put. "Bridgette's parents know — if you want to see us, you have to come to our house for Christmas," Jeff says.

"It's this big, damn deal every time," Bridgette says. "Jeff's mom winds up not talking to us every year."

"Don't even start me on that," Marie says, when I mention it.

The present-day revolution in the art of going much too far with your Christmas lights began, most everyone agrees, with Carson Williams, a forty-year-old father of two and electrical engineer who lives in Mason, Ohio, a middle-class suburb between Cincinnati and Dayton. In 2004, using software and equipment from a small New York–based manufacturer called Light-O-

Rama, Williams programmed the 16,000 lights he'd hung on his house and shrubs and trees to "dance" to an instrumental song, again "Wizards in Winter" by the Trans-Siberian Orchestra. (The Trans-Siberian Orchestra, aka TSO, bills itself as a progressive-symphonic rock orchestra and has recorded three hit Christmas albums since 1996, tapping the same market niche enjoyed by the multi-platinum-selling Mannheim Steamroller. "Wizards in Winter," from the album *The Lost Christmas Eve*, is the "Stairway to Heaven" for the men of America who put tens of thousands of Christmas lights on their suburban houses and program them to blink to music.)

Williams was not the first to have this idea — for several years Light-O-Rama and other display suppliers had been selling such equipment and software to homeowners, including Jeff Trykoski — but his display hit a sweet spot and drew the most attention, mostly because he posted a video of his house on the Internet. Hundreds of people began turning up each night. He'd rigged up the music to play on a short-distance radio station, so cars could tune it in, park for the entire song, and then move on. The following year, in 2005, Williams added another 10,000 lights to his display and posted a new video, and that's when it became a worldwide phenomenon. The video got more than a million hits. Miller beer made a TV commercial featuring the Williams house; the *Today* show and other national media came to interview him.

At that point, Jeff Trykoski became the town wonder of Frisco for a similar display, even though he never quite achieved Williams's instant fame. Jeff did win a nationwide lights display contest sponsored by a website called Planet Christmas and, like Williams, got several hundred thousand hits on a YouTube clip of his house. Lights nerds from all over the country e-mailed Jeff for advice, and he gave them his blessing to copy his designs. Online, Jeff and several others formed their own Texas

Christmas Lights Club, for which Jeff serves on the unelected council of poobahs. Jeff saw the entrepreneurial potential here and started his own consulting business on the side, called Illumimax LLC. As his mother says, *It could never just be a lemonade stand.*

A year after Jeff won the Planet Christmas prize, the lights are still not bright enough — not big enough. This is why Jeff signed a contract in the spring of 2006 to oversee and light up a new "mixed-use" retail/office/luxury-apartment development called Frisco Square, which will require 150,000 lights at a minimum (or about three times the number on his house), programmed to blink and dance to four different Christmas songs. The goal is to get 100,000 people to drive by and gawk. "They've taken my husband," Bridgette says, resigned to it. "I won't see him until January."

To the 60,000 or so people who've moved to Frisco since 2000, the city's downtown Main Street recalls the decades before the new affluence: it is a shabby strip of nail salons, a soft-serve custard stand, a now-beloved water tower first built by the St. Louis & San Francisco Railroad in the early 1900s, and nearly empty grain silos (no one farms anymore), backed by neighborhoods where Latino immigrants have found cheaper housing near a gas station where the day laborers wait for work.

For fifteen years, the city of Frisco's recreation department has sponsored a "Merry Main Street" downtown on a Saturday night in early December, with volunteer help from a loose collection of small-business merchants and garden club members. The event featured a tree-lighting ceremony, some caroling performances by schoolchildren, and family-oriented activities such as face painting and horse-drawn buggy rides. There was also a live-action manger scene staged by the First Baptist Church at a gazebo next to an insurance office. It had been

homey and cheap, but also genuine — the very small-town quality that newcomers, in survey after survey, said drew them to
buy one of the thousands of new houses being built in Frisco.

Now the city has grown so large — and its budget has like-
wise swelled — that ambitions for an annual Christmas event
have grown in scope. Merry Main Street 2006, city leaders decided, will this year move a mile west to the Frisco Square development that includes a new $31 million City Hall, which has
a gleaming limestone and granite façade and a retro-style clock
tower.

The basic look and scheme of Frisco Square can be found in
any exurb or downtown revitalization project of the past decade, inspired in theory (if not practice) by the post-megamall,
New Urbanism movement. The goal is to get people to park
their cars and turn, briefly, into consumer-pedestrians; to per-
suade those with the most disposable income and the least occupied uteruses to live in apartments closer to where they can
shop, work, exercise, and be entertained. Frisco Square's management team openly covets the success of the Shops at Legacy,
located eight miles due south in Plano. The Shops at Legacy is
a thriving collection of condos, restaurants, sports bars, martini
lounges, a trendy-indie movie theater, boutiques, a hotel, and
office buildings. These places are all the rage in the booming
2000s; by the winter of 2006–7, there are no fewer than five
developers unveiling their pastel-hued, computer-generated ar-
chitectural renderings of mixed-used projects in Frisco, planned
to be built along the new Dallas North Tollway exits under construction.

The master plan of such places always involves a village
square setting. Music is piped in from small speakers in every
tree and planter, an inoffensive blend of Rat Pack and Ella
Fitzgerald standards, or adult contemporary hits. It's a stroll
through a Norah Jones and Maroon 5 fantasia, a world exist-

ing on the precept that Americans need, at all times, places to buy $300 bed linens and order $14 cocktails. You'll know you're there by the faux period architecture. You'll know it when you see the luscious photomural montage of fresh yellow, red, and orange bell peppers symmetrically piled atop one another in a "farmer's market" crate.

Advertising images for these places are always shots of ethnically blended groups of people buying organic groceries, or having one another over for cocktails in their loft condos, where they all sit on West Elm furniture and admire the Photoshopped view (from floor-to-ceiling windows) of a twinkling city that isn't ever the city you're in. Later, in another advertising photo, these mixed-users go to the charmingly hip coffeehouse, where they sip from mugs and check their e-mail and stock investments via the free Wi-Fi access. After which, they are seen walking Boston terriers and yellow Labradors in parks that are supposed to be nearby. Such parks do exist: hills carved by groomed hike-and-bike trails, where sod has been laid and the trees are at first barely taller than you.

With the approach of Christmas 2006, Frisco Square consists of just three buildings, not including City Hall. The architecture features Italianate columns and arches over balcony terraces, encouraging the shopper-diner-citizen to dial back a century. It is ready for people, yet the people do not seem ready to flock to it. At least half the retail spaces are empty, but there are three restaurants (including a Subway sandwich shop), a chic day spa, and a high-end overstuffed-furniture gallery called Bella Cosa. The vacant storefront windows are covered in plywood, onto which are painted long, colorful murals of leisurely happiness at a European-style fresh-food market, where people shop and stroll. Above the retail level of both the east and west buildings are five floors of office space and rental apartments.

The apartments, I am told, are 90 to 95 percent occupied, with rents around $1,200 a month, average.

The only person I meet who is living the complete vision of what the developers have in mind is Matt Lafata, a thirty-eight-year-old city councilman who works at a human resources consulting firm called HRchitect, which keeps offices in Frisco Square. Matt, who is planning to run for mayor in 2008 (unsuccessfully, it would turn out), also lives in Frisco Square, in a two-bedroom apartment with a nice balcony view of the usually empty sidewalks below. He walks both to his job and to city council meetings. Matt has a fiancée, Erika Howe, a former beauty pageant winner who sells advertising for Jack-FM, a big radio station. Not so long ago, Matt and Erika were neighbors in a Frisco subdivision called the Plantation Resort. They fell for each other, and now, with six kids between them and divorces behind them, they are looking at the future as a Starbucks-era Brady Bunch. Matt and Erika like to have parties at Matt's apartment, inviting all their friends. At their wine-tasting party in the fall, it occurs to me that we are as close as one can get to the advertisements of the heavenly, happy life in the new mixed-use America.

What Frisco Square needs to perk things up is a little Christmas, which in management's eyes means a whole lotta Christmas. The Shops at Legacy is already famous for a giant "Lights at Legacy" night in November each year, which has been drawing crowds of 20,000 people to watch its six-story-tall tree be lit by Santa Claus. To compete, Frisco Square wants something bigger but different, and this is what leads them to lure Jeff Trykoski, at first for advice. Could the same kind of show he stages at his house on Bryson Drive be replicated on the Frisco Square buildings?

On hot spring and summer days, Jeff began having meetings with people from Frisco Square and City Hall to envision a synchronized light display for all the buildings, at least 150,000 bulbs. These would dance in time to a selection of four Christmas carols, broadcast on a low-frequency FM signal and also piped through the retail area's outdoor sound system, all controlled from one computer, which Jeff would program. Frisco Square signed Illumimax, Jeff's one-man company, to a five-figure contract. (Jeff asks that I not reveal the exact amount he's being paid, but he will say that it is going to cover a couple of semesters of tuition for the MBA degree program he is to begin in January 2007 at Southern Methodist University's commuter campus.)

At the same time the City of Frisco also hires Jeff to design the Christmas lights for its nice, new City Hall building, which had just been completed on the south end of Frisco Square. (Jeff's contract for that part of the project is an additional several thousand dollars, paid by the city, and it's a matter of public record should anyone wish to look it up.) Frisco Square and the city will pay their own work crews to install all the matching lights on the buildings and the trees. The city and Frisco Square will also buy all the lights, extension cords, wiring, wireless Internet transmitters, and, of course, the eighty Light-O-Rama circuit control boards that will control the show.

Jeff spends most of the year confidently designing the holiday display, working from architect's drawings of the buildings, some of which are still under construction. He decides which bulbs will be white, which will be red, where there will be bell-shaped lights, and where there will be giant snowflake lights. Some evenings, after he gets off work, Jeff climbs, alone and untethered, onto the roof of City Hall to take measurements, or test Wi-Fi signals, or just be Batman on a parapet. He becomes intimately familiar with the location of electrical outlets on each roof. If it's possible to fall in love with office and apartment buildings, Jeff does. He dreams about them, dreams in which something always goes wrong and has to be fixed by Jeff alone. He orders several pallets of C-9 clear bulbs, and dozens of snowflakes and bells for the accent lights. He buys 200 extension cords at a Home Depot grand opening sale. Though he has no personal investment in the shopping center, Jeff takes this mission seriously. His city needs him. Here is a chance to plug a lot of stuff in at once and flip a switch. He doesn't want to tell anyone this, but he'd probably do it for free, just for a chance to see all that electricity zap on at once.

Jeff doesn't work completely alone; he has a secret weapon. For a small cut of the action and the price of keeping his refrigerator stocked with beer, Jeff enjoys the musical aptitude and assistance of his brother Greg, who will sit at Jeff's home office computer many nights in a row in October and November, put on a set of headphones, and painstakingly program the songs that will accompany Frisco Square's light show. It was Greg who figured out, years ago, how to precisely choreograph Jeff's house lights. Jeff does the math and electrical schematics, but Greg does the art. Note by note, circuit by circuit, Greg needs at least fifteen hours to program one song across a spreadsheet of grids on two computer monitors. "It's easy, really," says Greg, who has played drums on and off since he was a teenager. "If you can keep a beat, you can do it. It takes forever and you will want to kill yourself before it's over, but other than that, no problem."

Greg comes over after work, often after a stop at Hooters to eat, pulling up in his black Dodge Ram pickup. He's wearing a "Got Freedom?" T-shirt. "I gotta go," he announces, heading straight for the bathroom, which Bridgette has just decorated with snowman guest towels and a snowman-motif shower curtain and a matching snowman soap dispenser. ("Nice," she says, the longer he's in there.)

After a while, he emerges and trades a few obligatory barbs with Bridgette, as is their habit. Greg lived in Jeff and Bridgette's guest bedroom for a year while he wasn't working and was up to his goatee in bills to pay. Bridgette used to ask him why he just couldn't get a job at Home Depot, "or do something," she'd gripe. Just when they truly couldn't stand each other, Greg found a new job, moved to his own apartment a few miles away, and a truce was declared. Now they've reverted to a more comfortable level of petty but affectionate jabs. He is allowed to snark off to her on any subject except two: No jokes about her enhanced breasts, even if she makes one first. Also, no jokes about inventive ways to kill her cat.

He could not be more different from Jeff. Greg has the charming, urban-redneck bachelor act down pat, always on, a walking YouTube clip of comic timing and near-perfect routines from *Talladega Nights: The Ballad of Ricky Bobby*. They don't even look like brothers. Greg works for a division of Exxon-Mobil, performing something called lubrication failure analysis. "I'm a certified lubrication specialist," he says. (Get it? Get it?)

Greg eventually settles in front of the computer, working until he's about to pass out, sometime after 3 A.M. The big celebratory turn-on at Merry Main Street is only a few days away. The lead song — as agreed upon by Jeff, Greg, and Bridgette — is Mariah Carey's 1996 holiday hit, "All I Want for Christmas Is You," which will be followed by Trans-Siberian Orchestra's "Christmas Eve in Sarajevo." Then will come Mannheim Steamroller's "The Little Drummer Boy." Bridgette also insisted that the brothers use Amy Grant's "Mr. Santa" (based on "Mr. Sandman" — "Mr. Santa / Bring us a dream . . .") because she thinks "The Little Drummer Boy" is too slow, too boring.

These all started out as perfectly fine Christmas songs. Who in his right mind doesn't love Mariah's trilling? "I don't want a lot for Christmas / There is just one thing I neeeeed . . ." But the more I'm around the Trykoskis, the fewer conscious moments I have when "Wizards in Winter" and the other four songs are not playing over and over in my brain, the Mariah Carey one especially. Greg tells me that eventually you get over it, but not without getting good and sick of Christmas first. "You start to hate everything about Christmas," he says, "and then we finish and you have just enough time to start to like Christmas again." A calm despair begins to settle in the thousandth time I hear "All I Want for Christmas." Leaping off the top of Frisco Square is not out of the question now, if it will make the Mariah stop.

5

Anthropologie

THE SMOTHERING SUMMER heat at last relents to a North Texas autumn, with temperatures in the low seventies. The sky above Frisco turns papery white some days, a dishwater-colored screen waiting for a holiday-themed movie to start. Starbucks brings out the red cups ("Cheer Gathers" is the slogan this holiday), and the moms glide in wearing boas and Uggs, trailed by their princess-daughters, and they both order grande mocha lattes. Men spend extra time in the Bass Pro Shop megastore, looking at rifles and things camouflage. It is "winter" now.

The Plano school board hosts an "informational" public meeting on a Monday night in early November, featuring a free-speech attorney named John Ferguson, who is also a Baptist minister. He is here to tell an audience of 200 about celebrating Christmas in public school classrooms. He carefully walks us through his PowerPoint presentation of the "three Rs" — rights, respect, and responsibility. The awareness seminar has become a seasonal legal necessity, after a rash of culture-war skirmishes in the Plano school system made national news several times between 2001 and 2004, over such matters as classroom goody bags distributed with "Jesus Is the Reason for the Season" pencils and candy canes attached to a little story alleging that the iconic treat is a *J* for *Jesus,* stained with red stripes of his sacrificial blood—yum! "It's the most litigious time of the year," sang the first line of a story about the forum in the *Dallas Morning News.* The free-speech attorney answers only preapproved questions the audience has submitted on little blue cards: *Are schools allowed to decorate trees? May students and staff wear red and green? Are you allowed to wish a staff member or a student a "Merry Christmas" while on school property?* (Yes, yes, and probably yes — but let's go back over the three Rs, shall we?)

After last call one night at the Applebee's bar at the Centre at Preston Ridge, I walk out to my car and am drawn to the sound of a man in a cherry picker shouting in Spanish, and a woman's voice near the ground, shouting back at him. They and three other coworkers are putting up a twenty-five-foot-tall Christmas tree.

The modern-day Christmas elves here are almost all Spanish-speaking immigrants. They can earn around $10 an hour, depending on the contractor. They never leave a trace. I see various crews of Latino elves working late in the lobby at a nearby office park. Another night they come and put wreaths as big around as hot tubs on a Blockbuster Video store. One night they unpack a herd of lighted reindeer and place them in random grazing order on a grass median. At another strip mall, they place giant red and green gift-wrapped "presents" on the knolls between parking areas. They wrap oak trees in clear lights, all the way up the trunks and meticulously around branches, working until dawn. They laugh and trade insults. They curse at the night wind when it knocks over a glowing plastic snow globe containing a penguin in a Santa hat.

* * *

The mythological gesture that gave birth to this land was a transaction. A transaction is the most mediocre form of human intercourse. It is an exchange of mutual fictions rather than of true feeling. It is a way for two people to take at the same time, together.

— TIMOTHY "SPEED" LEVITCH, "Wall Street: The Story of
What Happened to Our Intimacy"

I now get the lay of the land. As a child of the suburbs (and now as a man unashamed to be seen closing the bar at an Applebee's), I suppose I always got it. The best way to navigate America is not as an explorer but as a consumer. I put away my books on North Texas pioneer history. I stop reading about the local soil strata and a century's worth of drought trends. I stop thinking of places and life in relation to U.S. Geological Survey maps and Google Earth and instead think about life in relation to the strip centers and the malls. What was once geography becomes mallography, where you identify yourself more by your box-store preference than anything else. (Target person? Wal-Mart person? One is hardly ever both.) You draw lines in relation to the malls you most like, describing yourself and your surroundings as part of an evolving autobiography of where you shop. *Coming Soon* is the grammar of the place. To an outsider, it's all one featureless, never-ending mall world.

I spend part of every single day in Frisco's Stonebriar Centre mall, among its five anchor stores and 145 shops and restaurants. The clothing stores all play carefully chosen songs that are girly, spangly, designer-handbaggy, the songstresses chirping about their sudden flashes of soulful insight and independence: *Suddenly I see, this is what I wanna be.* Lyrics that go, *Feel the rain on your skin . . . Only you can let it in . . .* Then slacker troubadours like Jack Johnson or John Mayer swoop in, with their surfy love ballads and their dreamy power to open purses and wallets.

People sometimes go to the mall twice a day. On weekday mornings and afternoons it is the "Strollerbriar" of its nickname, filled with bored moms who visit over and over again, eddying out by the play area to watch carefully as their children maniacally romp on toys in the shapes of a giant, smiling plastic cell phone and a computer terminal. On Friday nights Stonebriar Centre fills with packs of teenagers, who seem to have stepped right out of television shows about teenagers, who screech joyfully at one another between checking their phones. Emo rocker teens with pink-tinged shag haircuts and Joey Ramone drainpipe jeans gather at tables by Sbarro pizza. On Saturday nights there are married couples, MILFs with their DILFs, who've hired babysitters so they can have dinner at the Cheesecake Factory or California Pizza Kitchen and now wander around Barnes & Noble, browsing together and then drifting apart, until it is time to ride the escalator together up to the AMC 24 for a 9:20 showing of a comedy starring Will Ferrell or Will Smith or Will Anybody. On Sundays, Stonebriar fills with football widows who paw lackadaisically through the sales racks at Macy's and Nordstrom. On weeknights, near closing, lonely employees stare abjectly from the Brookstone and Hot Topic and T-Mobile.

I like it *all*, in spite of myself. I spend hours traversing it, watching life in the Gap and Gap Kids, snooping around in J. Crew, learning things in the Bose stereo shop. I get all my gossip from guys who work at cell phone kiosks. I reward myself every few days with a $20 chair massage from the Chinese masseurs, who work in the concourse across from J. C. Penney, by the Dick's Sporting Goods.

Entire academic careers have been given over to trying to

understand what it is, exactly, this love affair between the mall and us. There are consultants, experts, antagonists, and a few poets. Very often the mall is approached as a problem, a symptom of social dysfunction, an abomination to the spirit. Not once have I met anyone in Frisco who complains about the surfeit of chain retail choices here, with easy parking and endless chances to shop a lot or shop a little. They are not immune to sudden (often faith-based) realizations that shopping can't buy eternal salvation, but these worries tend to pass quickly.

Here, chain stores are not merely redundant: chains bind us together, make us familiar to one another no matter what state we live in, give us comfort as they take away a sense of the local. The features that suburban anthropologists find bland and predictable have, around Dallas and Fort Worth, the eerie attribute of actually being local, starting with the area's most iconic area employers: American Airlines, Southwest Airlines, Exxon-Mobil, Neiman Marcus, and H. Ross Perot's Electronic Data Systems, now part of Hewlett-Packard, are all based here. Many familiar logos of the mallscape have headquarters here, too. In 2006 that list includes Blockbuster Video, J. C. Penney, FedEx Kinko's, Pier 1, Radio Shack, CompUSA, Michaels arts-and-crafts stores, the Container Store, the Bombay Company, GameStop, Tuesday Morning, and Zales, to name a few. Much of America's snacking begins here, at the corporate headquarters of Pepsi, Frito-Lay, Cadbury-Schweppes, and Dr Pepper; 7-Eleven and Pizza Hut are based here, as is a bounty of casual dining chains, including Bennigan's, Chuck E. Cheese's, and the company that owns Romano's Macaroni Grill, On the Border Mexican Grill, Maggiano's Little Italy, and Chili's. (I once observed a man in Frisco get so angry over the discontinuation of fried cheese sticks after a menu makeover at Chili's that he informed his waitress that he would be speaking with his neighbor, who, he threateningly added, "is very high up there at Chili's.")

In zombie movies, the last surviving humans find refuge in the mall. In Christmas movies, the mall mayhem triggers the protagonist's profound insight that none of it matters so much as love, and love is never found in a crowded mall in December. I don't quite buy that.

In the acres and acres of strip-mall and box-store panorama encircling Stonebriar Centre, you have a prime example of the very thing derided in the canon of "smart growth" and other twenty-first-century suburban planning movements: the obliterating effect of the megamall, that limitless supply of the sense of no-place. According to filmmakers, hipster television dramas, and countless depressing coming-of-age novels, this is where you're supposed to find America most imbued with a vague unhappiness, a shallowness, the cheating hearts and the hot tubs, the desperate housewives and their desperate husbands — all victims of hypocritical values and vapid demographics.

Frisco's city manager, George Purefoy, drives me around town in his white pickup one fall afternoon, late in the day, and we look at and talk about all of it. Purefoy looks like Gary Cooper would, had he played an urban land zoning visionary. He is fifty-four years old, tall, and talks with a gentle drawl. He has been city manager since 1987 and is generally credited as the master hand that made possible all this development and the twentyfold population boom. Purefoy tells me he sometimes worries that even with all the brand-new sports facilities, parks, trails, and other family-oriented fare now in Frisco, the result is a city in danger of never getting acquainted with itself or its history. It could very well be a collection of consumers who do nothing but hibernate in big houses and emerge only to commute, shop, or attend youth soccer games. It takes a lot of work and thought to make a nowhere a somewhere. Building the shopping malls was the easy part, he says, by comparison.

The people who drew the street maps and zoning plans and made modern Frisco are fond of telling the legend of a bull-dozer. The details change a little in each version of the story told to me, but the essential mythology remains fixed in their lore: In the late 1980s, when fewer than 6,000 people lived here, someone hired a man to drive a bulldozer back and forth along the pasture on the Frisco side of the Plano/Frisco city line, with no particular purpose other than turning the dirt. People like to imagine the bulldozer as an act of faith and deception. In one version, the mayor was the one who hired the bulldozer. In another, the mayor drove it himself. (Neither is true, according to Bob Warren, who is eighty-three and was Frisco's mayor then. He remembers the bulldozer. Or, more exactly, he remembers the talk about it. Similar bulldozer stories, it turns out, are common apocrypha in many American exurbs that all wound up with big-box shopping centers.)

The 1.1-million-square-foot Stonebriar Centre opened in August 2000, on that same pasture that allegedly had been faux-dozed more than a decade earlier. People in Frisco remember a decade's worth of handshakes and conference rooms and steak dinners with out-of-state mall developers. They remember the color-coded maps, the right-of-way issues, the zoning ordinances; promises kept and broken. They now keep pictures in their offices of the mall's groundbreaking ceremony and ribbon-cutting celebration, with the same affection with which they display pictures of their kids. One city councilor tells me his favorite picture of all time is an aerial shot of Stonebriar and the strip malls that surround it, taken on a December Saturday afternoon, in which it appears that every available parking space is filled. He likes to look at this picture and imagine the beeps of cash registers.

It's the strangest sort of pride. There is one developer everyone still speaks of, the man who shook the most hands and got
the mall deal together. He was the one who never relented, even
when it seemed the owner, General Growth Properties, wanted
to build on the Plano side of the city limits instead. People keep
mentioning him to me, and I'd like to call and interview him.
"Unfortunately, you can't," Bob Warren tells me. About four
years ago, the man fell off a ladder propped against the roof of
his home and landed on his head. He survived, but never recovered. The city gave him an award not too long ago, and his wife
had him wheeled in for the brief ceremony, and no one was sure
if he recognized any of his old acquaintances.

I ask why he was up on his roof that day.

He was hanging up his Christmas lights.

Near closing time on the second Friday of November, as the
Stonebriar Centre security guards in Mountie hats are chasing
away the remaining teenagers and the retail clerks are doing
their ritual folk dance of the pulling down of the store gates,
Gary Cathey and his crew of elves are unloading two panel
trucks of boxes and equipment near the back end of Sears.

Some boxes are labeled "Swag Garland." Huge steel turntable
wheels and set pieces are unloaded and lined up along the wall
by the Children's Place clothing store and down toward the Tux-
edo Junction formalwear rental store. Wide rolls of plush red
carpet are carried in, along with a variety of giant sparkly present boxes with bows on them and yards upon yards of plastic
greenery flecked in white. Gary is the design director of a Dallas-
based company called Daryan Display. They do Christmas at
night, in brightly lit malls devoid of people.

"Jorge and Barry, come with me," Gary says. "Where's Luis?"

A man raises his hand.

"Is there another Luis?" Gary asks, suddenly confused.

"That *is* Luis," someone else says.

"I should have learned to speak Spanish twenty years ago," Gary says. "Where is Cisco?"

The crew numbers thirty-five tonight. The Jorges and Ciscos help move boxes of swag garlands into position from one end of the mall to the other. The union guys on tonight's job (they're the white ones, with delightfully demonic tattoos on their arms and calves) drive hydraulic cranes slowly into the mall, beeping all the way, to hang decorations from ceilings of varying heights. A team of Hispanic women will spend the night steam-ironing long red velvet drapes. Another team will do the grunt work of constructing Santa's Land, where Santa Claus will sit daily on a green upholstered throne in the middle of a toy workshop with a circus-cum-Mardi-Gras theme, around which are three small merry-go-rounds with stuffed teddy bears, jesters, and elves riding them.

Gary wishes I could see some other job his crew does, like the Morton H. Meyerson Symphony Center in downtown Dallas. They do some fabulous work in downtown office buildings. "This is a very commercial job," he says. "It's nice, but it's a shopping mall. It's average." In the center, past the food court, Stonebriar opens up into a five-story atrium, where the Daryan crew hangs six large, golden, upturned cones, each topped with a crown of fake poinsettias, spilling forth strings of clear lights twenty feet long.

Stonebriar Centre has had the same Christmas decorations for seven years now. Daryan Display designed the pieces, oversaw their manufacture, stores them all year in its warehouses, and comes and installs them each November — in one night — and then, just as stealthily, comes and takes them down in the first week of January. Gary estimates that Stonebriar Centre initially spent about a half-million dollars on its Christmas décor.

Gary is a friendly, bearded, fifty-year-old man in faded black jeans, his long hair braided and stuffed up under a black base-ball cap. He went to law school in Louisiana and dreamed of being governor, until he realized that gay men couldn't be out of the closet *and* be governor. He went to New York and got into set design, lived the Studio 54 party life, and when everyone be-gan to die of AIDS, he moved back home to the kountry, figur-ing he'd die, too, only he didn't, and this, in a roundabout way, led him to Christmas shopping mall displays.

By midnight, the red carpet of Santa's Land is unrolled and squared off, and the guys lift walls into place. Garlands are be-ing unfurled and shaped. The drapes are steamed and some have even been hung. Gary orders twenty pizzas from Domino's and several boxes of chicken wings. When the food arrives, these unlikely elves collapse on chairs and sofas in the concourse. All the ironing ladies get out their phones and start texting.

By five or six in the morning, they'll pack up and go. Gary's not sure shoppers will take much notice when they walk into the mall a few hours later. "It reminds me of something I read where Gloria Vanderbilt had been quoted. She said as a child she didn't realize that flowers wilted," he says, "because at night, when she was asleep, the old flowers were always replaced."

The mall echoes with the sound of the crane arms going up and then going down. To check on things, the boss will ride a bi-cycle from one end of Stonebriar to the other, looking for tilted swag. If Jorge and Luis don't get it right, Gary will likely fluff it himself.

The following Tuesday afternoon, ten days before Thanksgiv-ing, I'm walking aimlessly in Stonebriar again, when suddenly: "Sir, let me ask you, do you want to see something amazing, something that will make your holiday even better?"

His name is Eitan. He is standing at the Sears end of Stone-briar Centre, working at one of those kiosks in the mall concourse where they sell everything from cheap sunglasses to wireless phone plans to real live crabs with cartoon characters painted on their shells. The kiosk cart where Eitan works, next to a Dippin' Dots dessert stand, had been empty most of the fall. Today it is decorated and tricked out to sell holiday "Snow Powder."

Which is what, exactly?

"I will show you," Eitan says proudly, in a thick accent.

On the cart's display area, little plastic snowmen and reindeer figures dwell among fir trees on what looks like giant hills of cocaine. Eitan scoops some white, polymer-based powder into a clear plastic dish and pours bottled water onto it. The powder instantly soaks up the water and expands into a chilled pile of fake snow. "You can put it around your, uh, villages? Houses? Or sprinkle it around a train set, or whatever," he says. "It never dries up, it always feels fresh, and when you want it to be cold, you just add more water" — he adds more water — "and see! It becomes fresh again!"

It's $20 a box, but, he says, "today, special, buy one box get one free."

Eitan is twenty-three. He has curly copper hair, wears a maroon hoodie, and stands a little over six feet tall. He says he finished his compulsory service in the Israeli army — he had been stationed on the Gaza Strip — and then, like hundreds of Israeli youths before him, he answered an online help-wanted ad to work at an American shopping mall kiosk. In exchange for about eight weeks of working in malls, mostly in the South and Southwest (usually for six thirteen-hour days in a row, arriving before the mall opens and cleaning up and securing the kiosk when the mall closes), Eitan and the others receive free room and board from their employers. Once they're loosed from mall purgatory, many spend their wages traveling to beaches in Central and South America. Eitan tells me he and five others are living in a two-bedroom apartment off Preston Road in Plano, sharing a Chevrolet Cavalier to drive to and from Stonebriar. His new girlfriend, Tali, is also working at the Snow Powder booth. ("Hello," coos a friendly, sexy Tali, in low-rise jeans, running her hands through fake snow, dribbling it on a small Santa's sleigh and reindeer.)

The Israeli kiosk kids have become a familiar sight in malls everywhere, and they are an aggressive lot, practicing a salesmanship more at home in a Middle Eastern bazaar than in the anonymous comfort of American consumerville. (The guy at the Sprint booth calls them "the Israeli mall-cart mafia.") Here in the buckle of the Bible Belt, the Lord's chosen people are hawking the latest in hair-straightening irons. The men are mop-topped and sport a day's worth of stubble. The women glisten and often dance to the beat pulsing from the Abercrombie store. Most of the women wind up selling exotic nail buffers and hand lotions ("Miracle of the Dead Sea" or "Soapranos") and tirelessly beckon passersby all day long: "Miss, do you have quick moment?" they purr in broken English. "Sir, can I ask you quick question?" (They always want to see your hands and fingernails, to blow your mind with the "Sea-crets" of the Holy Land.) Shoppers learn to ignore them, but the next kiosk leads to more Israelis selling aromatherapy sachets or "Motion Pictures," which are gaudy LED-screen renderings of waterfalls and tropical beaches.

Eitan has been in Stonebriar only two days and in the United States for eight. He wonders if he'll lose his mind standing in the middle of the mall for thirteen hours a day, nearly every day for the next six weeks. The minute Christmas is over, he and Tali are flying to L.A., then Brazil, to hang out. He's trying to focus on that. "I've never seen Christmas," he says, looking down the length of the mostly empty mall concourse. "Is it really so big? People don't seem to be so into it."

Wait, I tell him. I buy three boxes of Snow Powder, which turn out to be six.

"Buy one get one free, today," Eitan reminds me.

I drop by the Snow Powder cart almost every day for the next three weeks, to say hello to Eitan and Tali and see how they're holding up, jot down some of their impressions, and see if the powder is selling as well as Eitan hoped. But after a while, Eitan is increasingly cagey about talking. He and Tali and the other kiosk kids won't give me their last names. Almost all the Israelis working in U.S. shopping malls do not have work visas, according to the *Forward*, a daily Jewish newspaper. Eitan's manager, Uzi, who is in charge of all the Israeli-staffed carts at three shopping malls, comes and shoos me away one afternoon.

For its entire life—seven Christmases and counting—Stonebriar Centre has enjoyed the services of the same "genuinely bearded," heavyset Santa Claus who brands himself as "The Big Guy." He works eight to twelve hours a day, seven days a week, depending on when the mall closes, which gets later as Christmas nears. (He takes three breaks lasting a half-hour each.) He arrives on the third Saturday of November and remains until six o'clock on Christmas Eve. Families in Frisco have come to know and expect this particular Santa. He has a genteel manner and is good-humored about even the most kicking-and-screaming child. He's also good with people's cats and dogs, which are brought to him to have their pictures taken on weekly "pet nights."

Steve Lay, the mall's general manager, claims not to know the real name of Stonebriar's Santa, or what state he lives in, or where he stays in Frisco during his mall gig. "He will not tell you, either," Lay proudly says. "You can try and ask, and he never breaks character." A genuinely bearded mall Santa is yet another status symbol: lesser, smaller malls out in the kountry have "bad" Santas with fake beards. High-end, suburban mall Santas generally sign contracts for anywhere between $10,000 and $50,000 per season. It isn't easy work, and many such men consider it a deeply important calling. A nationwide brethren of bearded Santas hold summertime conventions, to attend workshops on improving their performances and to vote in plenary sessions on bylaws and rules that guarantee common Santa standards. Some have agents.

At Stonebriar, Santa's arrival is greeted with screams and camera-phone paparazzi at 10 A.M. on Saturday, November 18. He is brought around the Nordstrom parking lot in the warm midmorning sunshine, atop the city of Frisco's vintage fire engine, the one used for town parades. Santa disembarks and walks into the mall toward the circus-themed Santa's Land constructed by Gary Cathey's midnight elves a week earlier. Kids are screaming for his hand. Parents struggle to get it all on digital video. Santa takes it slowly, pausing for handshakes and hugs. He works his way past the Zales.

Eitan leaves his post at the Snow Powder kiosk to get a closer look. "It's insane," he says. "I have never seen a Santa Claus. He is like Paris Hilton here." He waves to the bearded man.

Santa comes closer, catches my eye. Then he winks and points to me — silently, presciently. He is close enough now to touch.

Know this: As a newspaper entertainment journalist, I have stood on red carpets. I have talked to Meryl Streep and Jude Law and Kate Winslet on Oscar night. At parties, I've made small talk with Beyoncé and Helen Mirren and Jake Gyllenhaal. I have thought of something to say to Natalie Portman, Prince, Nicole Kidman, Halle Berry, and George Clooney. Now Santa seems to be peering into my soul, and I'm stricken, mute.

"It's incredible," Eitan says, returning to his kiosk, shaking his head.

6

Christmas Caroll

THANKSGIVING: IF NOT FOR Independence Day, it would be the last of the pure holidays, replete with the remaining vestiges of Norman Rockwell advertising and mad rushes of grocery spending. Slathered in upbeat, revisionist history (cheerful natives, robust colonists), Thanksgiving happily sidesteps the hyperactive retail machine and instead stakes its claim on reverence, retaining its aura of handmade construction-paper Pilgrim hats. It conveys a sense of national togetherness, pride, gluttonous helpings of iconic food items, and the moments we take to consider our blessings.

Then all hell breaks loose.

As soon as many families finish their turkey dinners and watch a football game or two on television, they are packing the SUVs and heading out for malls and box stores, to wait all night and doorbust Black Friday. In Allen, Texas, just to the east of Frisco, a large outlet mall decides to break protocol and open at midnight. Wal-Mart is opening its stores at 5 A.M., and so are Best Buy and Circuit City. Target opens at 6, as do most of the stores in Stonebriar Centre. The crowds begin forming Thursday in the early afternoon.

Caroll Cavazos makes a drive-through Taco Bell run on

Thanksgiving night, taking sacks of food to her nineteen-year-old son, Ryan, and his friends, who planted themselves in line hours earlier, to wait in front of Best Buy. It was only two weeks ago that Ryan camped with his friends for several nights in front of the Circuit City across the street, hoping to be among the first to buy Sony's new PlayStation 3 video game system at the moment of its release on November 11. (Ryan and his friends wound up buying several PlayStation consoles; they have a scheme to sell them on eBay at a markup, after months of pre-release hype, hoping to make several hundred dollars' profit.)

Caroll will come back with her ten-year-old daughter, Marissa, at a little after 4 A.M. Their plan is to watch Ryan get in first with the aggressors, and then Caroll and Marissa will follow inside, once everyone else is let in at the end of the line. Caroll has a Christmas list of gift items that, according to Best Buy's newspaper advertising circular (which leaked days early online), will go on sale only on Friday morning. She and her children treat Black Friday with both seriousness and giddy regard.

This is where I first meet them, keyed up in that sleepy dawn light.

I notice Marissa first, in her fuchsia jacket, twirling around with her Starbucks cup, hopped up on caffeine and interrupting her mother every few seconds. "Mommy, how many more minutes?" she asks.

"Marissa, calm down," Caroll says. "See this line? We have to wait until they're all inside."

"Mommy—"

"Marissa, hush," Caroll says.

Caroll has been checking every few minutes with Ryan, calling him on his phone. She sees him slip in, when the doors open right at five. As the last of the line of people are herded in, there's a push of other people from the parking lot. "Come with us," Marissa says. (*Come with us, and I do.*)

Black Friday makes ideal news footage, accompanied by the now clichéd, seasonal frets of "worried" and "anxious" retailers, who express this same worry and anxiety about the holiday shopping spree in almost any economic climate — good or bad, resilient or flat. According to the National Retail Federation, Americans make anywhere from 20 to 40 percent of all their annual purchases between Thanksgiving and New Year's Day. This includes clothes, accessories, toys, gadgets, gizmos, media, food, booze, décor, hardware, and appliances. In some retail categories, such as jewelry, Christmas can count for as much as 80 percent of the year's profits.

Post-Christmas stockholder disappointment can morph into Third World factory layoffs in a matter of weeks, even in marginally profitable years. The "worst" Christmas seasons in the past forty years still saw sales totals that were measurably higher than the prior year's — until the recession of 2008, when sales declined 2.8 percent, sent analysts and investors into a conniption, and furthered an economic swoon. (The best Christmas in the past ten years was 1999, when holiday sales increased 8.1 percent; in any case, the average increase between 1997 and 2006 was 4.76 percent. In that decade — arguably America's biggest and longest shopping binge — the total take on Christmas shopping grew from $300 billion per year to $456 billion.)

Best Buy has hired some 23,000 additional seasonal employees this Christmas season to help out at its U.S. stores, which now number more than 1,000. (During vigorous years, some 700,000 Americans are temporarily employed as Christmas help.) In after-hours training exercises for new recruits, Best Buy stages mini-riot dress rehearsals, teaching workers how to defuse arguments and calm more manic shoppers. Seasoned employees play the part of customers, cutting lines, shoving, grabbing merchandise from other shoppers, or accusing the store of hiding popular items in the back room. "Until you experience [Black Friday]," one training manager told the *Washington Post,* "people never grasp what's about to occur."

Caroll is forty-nine. She exercises and watches what she eats to keep trim and has her hair colored a deep brown. She's a single mother with two marriages behind her. She works as a team leader in the credit card division of a corporate branch of Bank of America. (Those low-interest refinance offers you get in the mail? The offers for affinity cards with your alma mater's mascot? The myriad come-ons for reward programs? Caroll's team —known as "files management"—helps make those happen.) She's worked in the credit card business for fifteen years now, first at the Dallas branch of MBNA America, which was once the nation's largest issuer of credit cards, until Bank of America bought it in 2005. Since then, she has worried where the consolidation of the two banks' tens of thousands of employees will ultimately leave her. So far she's played it cool, but she spends a lot of her time at work wondering how the teams will shuffle around, and if it's better to draw attention to herself and her projects, or better to keep her head down.

Her Christmas budget:

Caroll limits herself to spending $1,200 on Christmas presents in 2006. This is more than various surveys estimate the average shopper will spend—by about $400—but it is about half of what people earning more than $75,000 a year will spend on holiday gifts. Caroll fits right in the middle. Her budget lets her spend $300 for each of her three children, and another $300 for her mother and other friends and relatives. Much of it will be spent here at Best Buy this morning. During the year, Caroll has saved the cash she plans to spend this season, so when her own Bank of America Visa card statement arrives in January, she can pay off all or most of the holiday tab.

Her list:

• For Ryan Sullivan, her son, who will turn twenty years old in three weeks: Caroll has promised she'll buy him a Hewlett-Packard Pavilion D6000 laptop, but it has to count as a present for both his birthday, which is December 16, *and* his Christmas. The laptop is on sale, today only, for $479. "It's over my budget," Caroll says, and she's being firm this time, reminding Ryan that more stuff won't be magically waiting for him under the tree.

"It's funny," Caroll says. "Ryan actually works at this Best Buy." This is true. Even as he has spent the last thirteen hours sitting outside it, he must report back later today and clock in for an eight-hour shift *inside* it. Since his shift doesn't start until noon, he has to wait in line with the rest of the doorbuster crowd — store policy.

Ryan has worked for several months in Best Buy's "car-fi" department, selling car stereos, speaker systems, satellite radio packages, GPS devices, and such. He seems to like it, but Caroll worries about him. After graduating from high school, Ryan went for two semesters to Oral Roberts University, a Christian college in Tulsa, Oklahoma, five hours north of here. He wanted — and still wants — to study engineering, but he flunked calculus, lost his scholarship aid, and moved home to the safety of his bedroom and a diet of frozen pizza rolls. In addition to Best Buy, where he earns $10.50 an hour, he has worked before at a Home Depot, a J. C. Penney, a Blockbuster Video, and (his favorite job) as a parking valet. "I keep telling him it's his life," Caroll says. "And if he wants to go back to school, he has to do it himself."

Ryan either goes or doesn't go to the local community college courses for which he registered, Caroll says — she can't be entirely sure. A question rarely gets a direct answer. She keeps asking for a transcript, a course schedule, anything. Caroll keeps telling him that she can't make life happen for him, doesn't want to make decisions for him, doesn't want to keep bugging him or bailing him out of problems. She prays about it once in a while, and the answer that keeps coming back to her is to let him fall or succeed on his own. "He's a good kid," she says. "He's always been my sweetest child."

• For Michelle, her twenty-seven-year-old daughter, and Michelle's husband, Joey: "Michelle is my practical one," Caroll says of her oldest, the one with an independent streak. Michelle is finishing up her college degree and plans to teach elementary school. Joey, also twenty-seven, works at a Discount Tire auto shop, to which he'd returned after a stint in mortgage banking at Countrywide. The couple bought a house recently in McKinney, a suburb fifteen miles east of here. Michelle decorated the house in earth tones, modern furniture, "and really just the opposite of what I like," Caroll says. (Caroll likes bright colors, florals, decorative trim, window treatments.)

Michelle and Joey are expecting their first baby next June. It's Caroll's first grandchild. "Michelle has already bought everything. They've already painted the baby's room," Caroll complains, like a typical first-time grandmother. "I said, 'You're not leaving anything for anyone else to do.'" She sees becoming a grandmother as a way to maybe get closer to her daughter, who talks to Caroll nearly every day on the phone and yet still keeps her at a bit of a remove. This year, Caroll's present to Michelle and Joey is to help them pay off a new Whirlpool washer and dryer, on sale here today. Michelle has just arrived at the Best Buy parking lot, too, to meet her mother and make sure it's the right washer and dryer, and to set up a date to have them delivered. She stands with her arms crossed and regards me with skeptical blue eyes. (*He's a what — a writer? He wants to follow us around?*)

When Michelle was a teenager, she shut herself in her room and cried for what seemed like hours — over what, Caroll cannot quite remember. She came out and announced that she would never cry again in her whole life, and Caroll has not seen her cry since. "She is very tough," Caroll says. "She's like me, that way." They do have a mother-daughter ritual, on weekends. Joey usually works at the tire place on Saturdays, so Michelle meets Caroll and Marissa at the mall, usually Stonebriar Centre, and they walk around and shop and have lunch, maybe at the food court or one of the restaurants. Sometimes they'll be in the mall all day and never buy a thing. They wander around and talk about what they see.

• For Marissa, Caroll's third and youngest child — and constant companion: Nicknamed "Turbo" by her older brother, Marissa is in fourth grade at American Heritage, the same Christian school Ryan graduated from in 2005. Marissa wants *everything* for Christmas, or any day, and rattles off for me a random list of all she does not have but wants "so bad": a pink iPod Nano, a Nintendo DS with a game called Nintendogs, DVDs of *High School Musical* and *The Devil Wears Prada*, knockoff Coach and Louis Vuitton handbags, clothes from Abercrombie and Limited Too and Aeropostale. In the Best Buy parking lot, Marissa shows me the bejeweled phone she got for her birthday a couple of weeks ago, a gift from her big sister (over initial objections from Caroll, who still wonders why her fourth-grader needs a phone plan).

Most of all Marissa wants to be an actress, the next Disney Channel star. A year ago, Caroll enrolled Marissa in acting classes, which are held each weekend in a small theater in an office park not far from D/FW airport. Caroll drives Marissa there and waits in a theater seat or out in the car while Marissa's class acts out improv exercises and prepares to meet the Hollywood casting directors and agents who will one day come.

Marissa is the one family member who isn't getting anything from Best Buy this morning — "not that I *knowwwww* of," she emphasizes precociously, so her mother can hear. Caroll tells me later that she thinks this is the last year Marissa is going to believe (or pretend to believe) in Santa.

Caroll was almost forty when she found out she was pregnant with her youngest. It was her second marriage. She already had Ryan, who was nine, and Michelle, seventeen, from her first marriage. "And I'm asking God, *Why a baby?*" Caroll says. "*Why now?*" Caroll and her second husband split when Marissa was four. (They had married on a Thursday, and by the weekend, Caroll recalls, "I sort of already knew it wasn't going to work.") Although there is a joint custody agreement, it has been many months since Marissa last heard from her father.

Marissa was born a couple of weeks before Thanksgiving 1996, and Caroll believes now that God intended to send her a constant companion. They cuddle a lot — Marissa curls up to Caroll almost anywhere, whether it's at church or in a booth at Bennigan's. They also argue a natural amount, Caroll telling Marissa to pick something up off the floor for the fifth time, or insisting that no, they cannot have lunch at Which 'Wich on the way to acting class. Marissa listens in on most of Caroll's conversations. Marissa shows almost none of the social apprehension or shyness of her mother; she walks into a room and makes friends. Even as Caroll's world revolves around her daughter's activities and every need, and even as Marissa seems privy to every activity and conversation in her mother's life, Marissa will once in a while get out the boot brace she had to wear for a sprained ankle and limp around in it for a day or two. "She does that when she thinks she's not getting enough attention," Caroll says.

She wants to get Marissa a new bicycle for Christmas. Yes, there will be new outfits and cheap jewelry for her mini fashionista, but Caroll also thinks Marissa needs to go out and play more, while she's still a little girl.

- For everyone else: Caroll spends the remaining $300 of her budget on small items for friends, her mother, and her siblings and their families. She and one of her sisters started exchanging tree ornaments a few years ago, to cut down the hassle of trying to think of what to get each other. Caroll's mother (who lives three hours north in Oklahoma) has informed Caroll she wants a mink coat this year. "She actually told me to just send her money," Caroll marvels. Instead, this year, Caroll and one of her sisters are going to split the cost of a home computer for their mother — on sale for just under $300 at Best Buy this morning. Caroll is also replacing her own computer this morning, too, which will go in the desk nook off her kitchen. "It's just time for a new computer, and since they're on sale . . ."

That covers Caroll's list, but it does not cover Caroll. "Something I always thought was important when I was married was having someone to do for me," Caroll says. "Birthdays, Christmas, someone to just remember you." That never happens anymore. Although Caroll is always telling herself to seek out the new experiences around her, she also has a pragmatic, pessimistic streak. She tightens her jaw when she's trying to figure out what happens next, what the catch is. When I first saw her, Caroll reminded me of someone, a face I couldn't place. It took me a while to see it: in that Best Buy lot that morning, she looked like Florence Owens Thompson, the woman featured in "Migrant Mother," the famous 1936 Great Depression photograph taken by Dorothea Lange. It is that same taut, serious beauty, fixed on the uncertainty of whatever comes next.

When she was a girl, her nickname was Ricky. Caroll's father was in the air force and they moved a lot. They went to the

Japanese island of Okinawa when she was three. "Christmas, I think, is the only thing where I'm always sure the glass is half-full," she tells me. "I am still just like a kid about it, with wide eyes . . . I think it was the only time I was allowed to really *be* a kid." There is a Christmas morning so far back in her mind she almost can't remember it, at the house in Okinawa. In the middle of the night, Ricky got out of bed to go see if Santa had come. She followed the soft light to the living room and peeked around the corner.

She is not sure what she saw. It was a rocking horse, she thinks, seeming to shift and move with the blinking shadows from the lights on the tree. She thought it was a live reindeer. It scared her, and she ran back and told her brother. She hid under the covers. "I got scared when I saw the rocking horse and I thought Santy was gonna see me and I wouldn't get any presents." What did she see? The memory of that morning, that shadow, grips her with its simple and essential mystery.

One of Caroll's proudest accomplishments was making the final $637 payment on her house a while back, twenty-six years after she and her first husband signed the mortgage. The house is a simple four-bedroom late-seventies rambler in a once-rural suburb called the Colony, just west of Frisco. She stuck it out in this house, in good times and bad, and it's comfy; along the kitchen's floral wall trim, there's Caroll's collection of teacups, saucers, and plates. As the early twenty-first-century subdivision boom took off all around her, Caroll finally succumbed and starting flirting with a model home in a new development. It had the vaulted ceilings, the huge bathrooms and closets, and a breakfast nook at which Caroll imagined herself with some as-yet-undiscovered extra time in the mornings, sitting with a cup of coffee and gazing out the bay windows. The house — unbuilt, just a blueprint from the model — listed at $265,000. Right before Christmas last year, Caroll backed away from signing a contract.

"The way the housing market is going in Frisco, I could have turned around and sold it in three years," she says, which still holds true in 2006. Caroll is angry she hesitated. "Financially, I'm goin', 'Maybe that wasn't so smart.'" Then again, she never wants to lie awake at night worrying about every penny, the way she used to just a few years ago. Caroll watches her money carefully. She's had too many low points — leaky roofs, car repairs, interest rates, private school tuitions, medical bills, divorce. She has no problem imagining the world going under, even when everything's okay. She'll look at designer labels but never buy them. This sets her apart from most everyone I've met so far at the mall.

It takes sixteen minutes for Caroll, Ryan, and Marissa, now joined by Michelle, to have their Best Buy spree, navigate their way through the checkout line, wheel their cart to the door, have their merchandise and receipt inspected by a security guard, and be spat back out into the parking lot with one laptop, two PCs, and the soon-to-be-delivered washer and dryer. (Ryan's employee discount — 5 percent above cost — doesn't make much of a dent in the bill, since Best Buy marks many of the items below cost for Black Friday.) They also picked up a DVD player because it was on sale and Ryan convinced Caroll they might need it for some reason, even though they already have two. There goes almost all Caroll's Christmas budget. But, she says, they saved several hundred dollars over the usual prices. At Caroll's Taurus, Ryan says goodbye, takes his laptop, and leaves with his friends. Michelle's going home, back to bed. Caroll closes the trunk.

"Now, the next thing we always do is go to J. C. Penney and

get our free snow globe," she says. This started six years ago, when she and Marissa were in Stonebriar Centre on Black Friday with Caroll's mother, and Penney's was giving away small Mickey Mouse Christmas snow globes to the earliest customers. Caroll and Marissa now go back every Black Friday to get the new one. "We have one from every year," Marissa says proudly; she likes to line the globes up along the fireplace, next to the tree. After Penney's, mother and daughter spend the rest of the morning wandering the mall, Marissa asking for every last thing, and Caroll telling her *no, maybe, we'll see,* and *no.*

ShopperTrak RCT Corp., which monitors sales figures at almost 50,000 mall-based stores nationwide, reports total sales of $8.96 billion on Black Friday 2006, according to an Associated Press story released eighteen hours later, on what I have come to regard as Ennui Saturday. That's an increase of 6 percent over the previous year. These numbers arrive almost instantly, via an economic hocus-pocus rarely questioned by the business journalists who report the results. The numbers are then manipulated into various shapes and sizes, where the news peg is always essentially unchanged: *shoppers shopped more than they shopped last year.* The National Retail Federation announces that "more than 140 million people" — which would be 47 percent of all Americans — had, in one way or another, gone shopping on Black Friday. By one count, 28 percent of all purchases that day were things they had bought for themselves. "You don't go out and buy a flat-panel TV for someone else," Wayne Best, a senior vice president and analyst for Visa USA, tells the *Wall Street Journal* the following Monday.

The *Journal, New York Times, USA Today,* CNNMoney.com, Forbes.com, the *CBS Evening News with Katie Couric,* and every other media outlet traditionally portray the Black Friday sortie as a mind-blowing exercise in human desire. We are no longer citizens (or even just "people") as much as we are regarded as "consumers," the preferred term of economists and media analysts, who observe us like some species of animal that often but not always acts by market-tested instinct. Wall Street sees us as untamed and wild and dangerously fickle, capable of jeopardizing the economy with the snap of a purse.

Caroll and her family are like those everyday American shoppers you see on TV and read about in business-section stories, cornered for a moment outside department stores or in food courts by a reporter. There is nothing more anthropologically rewarding for economic soothsayers than to stalk the malls and ask real people how their shopping is going, how they feel about the economy this year, what their reactions are to prices on items they find welcoming (or repulsive), and what discounts they're responding to.

But on this Black Friday, I start to see it less and less as an objective act of Anthropologie. I see it the way Caroll sees it. Real lives are being lived here. People are shopping, but they are also falling in love, or kissing a child. They are sharing in a perception of glittery wealth. Many of them sincerely believe that the huge televisions they are clamoring for this morning will bring their families together in a more satisfying way. The restaurants, to me, no longer represent tastelessness. The plastic toys people seek for their children hold true wonder. In this carbed-out consumerismo are places and moments of true bonding, places to be seen and to see others, to simply exist.

As I'm feeling more warm and fuzzy toward the generic Whoville, I readily accept Caroll's offer to come with her to church and see what Pastor Keith has to say.

7

Unto Us

SEVERAL MONTHS BEFORE Stonebriar Centre mall first opened, Keith Craft and his wife, Sheila, announced that they were leaving their ministerial positions at an evangelical church in the Dallas suburb of Carrollton to start a new church in the undeveloped promised land of Frisco. Caroll decided to go along with the Crafts, because Frisco was closer to where she lives and she always liked Pastor Keith's sermons.

He named his new congregation Celebration Covenant Church, and the first worship service was held on a Sunday morning in January 2000 in a public elementary school cafeteria. A hundred or so people arrived early to set up the chairs and the music stage, and they stayed late to put it all away. Afterward, about thirty of them would go to Chuck's restaurant for burgers, and Pastor Keith would get the tab. He abolished the word *volunteer* and anointed everyone, including Caroll, as a *servant-leader.*

The church grew wildly. Caroll was (and still is) transfixed by Pastor Keith's trademarked message of personal and financial success through God. The pastor, like many church leaders before him, is an industry all his own, selling workshop seminars and DVDs to help disciples unlock the potential within, and he often travels the country to give motivational lectures to corporate leaders and employees, for which, he frequently mentions from the pulpit, he is paid $10,000 a pop. Caroll says his series of *Leadership Shapers* CDs taught her more than all the teamwork training she ever got at MBNA or Bank of America. "I was a high responder/low doer, which wasn't so good," she says, referring to the style of personality tests and metrics on which human resources departments and megachurch gurus thrive. Pastor Keith's charismatic teachings turned Caroll into "more of a high doer/low responder, or at least I'm trying to be more of one." For her it always comes down to Pastor Keith's most difficult truth: "God can show me, but he can't do it for me," she says.

Think, Be, Do is Pastor Keith's motto. He says it at every service at Celebration Covenant, at least once, and it is often writ large on the giant video screens at either end of the church stage. The Crafts' congregation now numbers 5,000 (a number that meets the general definition of *megachurch*), housed in their own "Celebration Center" building on a former pasture near a strip mall. Celebration Center has the approximate square footage of a Best Buy, but its current space is nothing compared with the $24 million cathedral complex Pastor Keith is raising money to build.

Caroll can usually be found in the second row at the second service on Sunday mornings, as a servant-leader in her chosen area of audiovisual technical assistance. She wears a headset and microphone so she can report sound or video glitches to the guys in the production booth or relay messages from the technical booth or backstage crew to Pastor Sheila or an assistant pastor in the front row. Caroll believes wholly in *Think, Be, Do*. It has not significantly increased her bank account or found her a good man, but she believes it's given her much more. "It's freeing," she says. "I'm just letting loose of fear when I'm there. I'm a very fear-based person. I've learned this. I'm hard on myself. I sometimes think that what could be happening for others is never going to be happening to me. Then Pastor Keith will do or say something that just makes me go, *Wait a minute* . . . It took me a long time to see that I was one of those people that the Lord was doing for. It's just this real humbling thing, how much the Lord really loves me and cares about me. Because I'm nobody."

You surely know what we're in for, if you've paid any attention at all to the rise of the megachurch and the new Christian culture's effects on American lifestyles and politics in the past two decades. You know what happens. The big-city reporter lands like an alien in flyover territory, walks in to the megachurch, and can't get over the fact that *they have their own Starbucks in the vestibule.* We're sure to be in for some holy rolling. I've been hitting at least one service at a different big (25,000 members) or small (200 members) Frisco or Plano church every weekend since September, and I can tell you they're irresistibly fascinating, trendy, skeevy, and ridiculous. To really get it, you have to go more than once. You have to go the next Sunday, and then another Sunday, and soon enough, something about them becomes endearingly earnest.

But before we go in and find a seat next to Caroll, you and I need to have a frank talk about Christmas. It's time. We must address two enormous myths about the American way of Yule.

One is the issue of history. What Christmas is and where it came from is a tangled mess of inaccurate origin stories mishandled on the freight journey from the Old World. The Puritans came to America in part to leave behind the very sort of godless travesties that Christmas represented to them, and they banned the holiday in much of New England, a prohibition that stuck for almost two centuries. The Christmas the early colonists fled was more like Halloween, when drunken street hooligans accosted good citizens in an annual shakedown for coins and booze. (That would be the "wassail"—it's not just a hot beverage, but a street crime.) "Traditional" American Christmas goes back a mere 175 years or so, courtesy of rich New Yorkers who became fixated, literally, on the better homes and gardens of Victorian England they were hearing and reading so much about. One of Manhattan's most prominent landowners, Clement Clarke Moore, is credited with penning the classic 1823 poem *A Visit from St. Nicholas* (also known as *'Twas the Night Before Christmas*), which served as a template for the modern version of Santa Claus. Santa is a mash-up of old myths that combine the legend of a Catholic saint with a host of wintertime night visitors, gremlins, and elves. In the 1800s, the middle and upper classes of New York began rejecting the street melee around them for a holiday centered on decorating the home and stoking the hearth. Instead of buying off street beggars in hopes of keeping the peace, the rich began rewarding the good behavior of children. In *The Battle for Christmas*, historian Stephen Nissenbaum notes that children of nineteenth-century America's middle class became metaphorical stand-ins for the unfortunates, whose proper behavior is rewarded with commercial goods that are mysteriously delivered by a kindly magic elf. Centuries of Christmas unruliness were overcome with lifestyle propaganda, the first "war on Christmas," fought by the Martha Stewarts of the day.

The sensational success of Charles Dickens's *A Christmas Carol* in 1843 furthered the complicated joy/guilt dynamic of the holiday and the notion that one could redeem oneself through a magnanimous surrender to the intangible idea of "Christmas spirit." Many of our hang-ups about Christmas — the stress and depression and psychological inadequacies — we owe in great part to Victorians on either side of the pond. But to their credit, they (and their pioneer marketing departments) gave us Santa Claus as we know and love him, and Christmas trees, and wrapping paper, and a bulk of the carols we still sing. They gave us most of the wonder, too: the idea of waking up to find new toys in the living room blew America's mind and never stopped blowing it.

We could fact-check the history of Christmas all you like — the "ancient traditions" that often trace back to nineteenth-century advertising more than they lead back to Teutonic folklore; we could revisit the holiday's role in class warfare through the ages; we could delve into the history of mass-produced toys or the rise of the American department store (and Santa Claus's codependent relationship with it). Much of what we consider to be "the true story" of Christmas was and is fudged for our enjoyment, and at any point you are free to respond: *So what?*

Then there's the other issue: theology. This is a high-voltage wire I hesitate to touch, for lack of wits and expertise, to say nothing of my lack of faith. Virgin birth and angel visits aren't the only big leaps here. A dip into even the more reverent inquiries by Bible scholars easily leads to the conclusion that there was no actual manger scene in Bethlehem, no shepherds dropping by to see the baby, no star in the east, no Magi, no frankincense, no myrrh. A Christ-child "infancy narrative" (as theologians refer to the Christmas story) appears briefly, and only in the Gospel according to Matthew and, more lyrically, in the Gospel according to Luke — both written around 80 A.D. The second chapter of Luke contains the passage famously read to us each year in *A Charlie Brown Christmas* by Linus van Pelt, who stands alone on the school auditorium stage ("Lights, please") to tell Charlie — beautifully, simply — the Gospel account from the King James version: "And, lo, the angel of the Lord came upon them, and the glory of the Lord shone round about them: and they were sore afraid. And the angel said unto them, Fear not: for, behold, I bring you good tidings of great joy, which shall be to all people. For unto you is born this day in the city of David a Saviour, which is Christ the Lord. And this shall be a sign unto you; Ye shall find the babe wrapped in swaddling clothes, lying in a manger . . ."

In fundamentalist churches in America, where the Bible is literally taken at its divine word and serves mainly as a springboard for encouraging congregations to strengthen their marriages and careers and vote to the right of center, the infancy narrative of Christ is second only in popularity to the Passion (the Crucifixion and Resurrection of Jesus), and there is no urge (nor is one welcome) to parse the Nativity story for historical veracity. Nevertheless, many scholars have concluded, some more gently than others, that the Christmas story is intentionally fictive, written by the earliest, first-century evangelists to beef up Jesus's street cred as a believable Jewish Messiah. Like any superhero, Christ needed an origin story rife with the drama, metaphors, and meaningful symbols of the era.

A theologian and priest named Raymond E. Brown (no relation to Charlie) wrote an exegesis of Matthew's and Luke's Nativity passages in his 1977 book *The Birth of the Messiah*, taking on the mythic story of Christmas as it aligns with the original Gospel texts. Brown was among the first scholars to try to determine the motivations of the first-century writers who invented the Nativity, and his work still stands as definitive among academics. As a man of faith, he saw that scholarship could not (and, in his view, should not) dent a deeply held devotion to the story of the manger baby. Until his work unleashed a rash of Nativity studies by others, Brown worried that most scholars had seen the biblical Christmas passages merely as "folklore, devoid of real theology . . . fit only for the romantic and naïve. As a result the infancy narratives are often overlooked or treated cursorily in seminary courses, even though those ordained to parish ministry will have to face them every Christmas."

Theologians, Brown wrote, "generally give [the infancy narratives] short shrift, disproportionate to their role in Christian theology, art, and poetic imagination." But questioning the accuracy of the Nativity should not deter one's faith in it, Brown believed: "Indeed, from this point of view, the infancy narratives are not an embarrassment but a masterpiece," he wrote.

Christmas first showed up on Roman Catholic liturgical calendars in the fourth century, which church historians see as an attempt to fuse religiosity onto an already entrenched bacchanal, a December 25 compromise for a feast day to celebrate the Lord's birth in a week everyone was solstice-partying already. Which is a long way of saying that Jesus *isn't* the reason for the season, and neither, really, is St. Nicholas or whatever you think our European ancestors called him. December's darkest days were always about consuming, feasting, reveling, and bingeing. Seen this way, our shop-till-you-drop impulse at the end of the calendar year is an ancient part of who we are. The shopping — the modern equivalent of harvesting and feasting — is perhaps more historically appropriate than the praying and believing.

There, I said it.

But where does that get me?

Not any closer to Caroll and her belief in Jesus Christ — manger baby and savior to all — and that's where I'd like to go. Church is far more important to her than Best Buy.

The male servant-leaders of Celebration Covenant take special pride in directing church traffic into the parking lot. There are a dozen of them in orange vests waving orange flags, chests puffed out, pompously motioning you around the worship center and into your parking space. Walking in, you get a warm assault of greetings and smiles, doors held open for you, your hand shaken repeatedly by still more servant-leaders. (And yes, there's a gourmet coffee shop in the vestibule, called Master's Blend.)

In the worship center, the service kicks off with a set list that is almost always the same — two rafter rockers and a power ballad. The Celebration Covenant praise band can expand or contract in size according to the vibe of the Word and the message that particular weekend. Some Sundays, it's jacked up to Spinal Tap's level eleven, with a twenty-five-person choir, two percussionists, a multiperson guitarmy, and two keyboardists. Other weeks, they take it down, "Unplugged"-style, with mood lighting. Usually there are four or five lead singers, often including Pastor Keith and Sheila Craft's younger daughter, Whitney, who has just turned fifteen. Whitney was among the thousands of young hopefuls who turned out in downtown Dallas to audition for *American Idol*. (Try again another year, Paula Abdul encouraged her, according to Caroll, who heard about it at church.) Another star on the Celebration Covenant stage is Clay Jones, a nineteen-year-old guitar player and singer who looks as if he walked straight off *The OC* or *One Tree Hill;* he's the boyfriend of the Crafts' other daughter, Keela. (Quick, go listen to Whitney's or Keela's 'or Clay's demo songs on their MySpace pages. I'll wait.)

To attend Celebration Covenant is to receive constant updates on the Craft family's daily lives and many blessings. In just a few services, you'll be familiar with their nicknames. The oldest, Joshua (aka "Champion"), is usually away, studying for the ministry at Oral Roberts University, the school where Caroll's son, Ryan, went for two semesters — a fact not lost on Caroll every time her pastor tells the congregation another story of Champion's successes at school and in life. Keela, the middle child (aka "K-Diva"), loves to shop and is about to graduate from high school. Whitney's nickname is "Giggles." Pastor

Keith's stories about his children provide sudden, if sometimes incongruous, entry points to all sorts of life lessons in his sermons. The Crafts have been recording and selling cassettes and CDs of the family's Christian pop songs since the 1990s, including a Christmas album featuring the children. Pastor Keith's showmanship dates back to the late 1980s, when he founded Strike Force, a traveling troupe of power-lifting evangelists. Keith and other musclemen in mullet haircuts would dress in matching tank tops. Their veins bulging with the Word of God, they'd lift hundreds of pounds of free weights for the crowd.

The unbeliever can be forgiven for feeling as though the faith business is nothing more than show business. From the largest churches to fledgling congregations, the new Protestant churches of Collin County comprise the largest music, theater, and pop-culture scene north of the LBJ. Most ape the same Christian rock sound that is neither U2 nor Coldplay nor Pearl Jam nor Good Charlotte, but owes a significant debt to those and almost anything else on secular radio, especially the oeuvre of pop and country female vocalists. No fashion trend escapes the good shoppers of Celebration Covenant. The young women here are all Jessica Simpson or Rihanna; the older ones approximate the cast of *Desperate Housewives*. The men are all Tim McGraw or Clay Aiken or that guy from Creed — except for a vocally gifted associate pastor named Ray Harmon, who bears a more handsome resemblance to Chris Rock. The women are always wearing the latest mall fashions — boob-prominent dresses or tops, tights with heels (as seen on starlets just now in *Us Weekly*), or wispily crinkled skirts, or knee-high riding boots. Their hair is styled in soft TRESsemé curls in L'Oréal hues, or flat-ironed and glistening by holy intervention. The men are all studiously rumpled, hair carefully tousled with salon product and faces that are exactly stubbled and sideburned. They prefer wearing unconstructed blazers stenciled and embroidered with a combination of romantic vampire touches: griffins, royal crests, gothic crosses, and vine-like paisley swirls.

Everyone — soloists, chorus members, and most of the congregation — sings with eyes closed, squinting in ecstasy. Every ballad is about a deep love affair with God: "Wrap me in Your arms" swoons one song, and "feel His strength" goes another, and finally, "Lord, hold me."

It all sounds like ringtones for Christ. I resist the urge to make a run for the doors the minute it starts. In many ways, these churches are all the same: They all have the praise band fronted by a dude who thinks he's the Bono of Plano. They all have the complimentary gourmet coffees and valet parking for newcomers. They all have a pastor whom I've collectively nicknamed the Reverend True Religion Jeans, and he always tells Venus-and-Mars-style jokes about women and men and relationships, with props. (*Don't you hate it when your wife puts the toilet paper on the roll backwards? Don't you just sit there and say, 'Help me, Lord'?*) The Reverend True Religion Jeans always has a pretty wife. Whether his congregation is 1,000-plus or just 120 people that particular Sunday morning, the Reverend True Religion Jeans always insists something uniquely divine is happening right here, this minute, right here in. This. Very. Room.

But Celebration Covenant delivers it all with the most oomph. At forty-six, Pastor Keith still has his bodybuilder's torso and arms and a shock-jock vibe; Pastor Sheila has giant eyes, teased hair, and the silky attributes of a precious teacup Yorkie. She dresses like that sexy real-estate agent with the month's highest home sales, in lots of leopard prints and cinched belts around her tiny waist — a Texas interpretation of the Annette Bening character from *American Beauty*.

In his sermons, Pastor Keith works himself up into a fever

over the "supernatural correlations" to personal success. He speaks in one meaningfully meaningless platitude after another. To wit: "Here's how to wreck your life," he'll shout. "*Don't* be the best of who you are. Do what you did in your last marriage, and you'll wreck this one. Do what you did in your last job and you'll wreck this job. Keep your same behaviors, keep your same attitude and whatever you have, do what you did the last time. When you think noble you are free. Free to be generous. Whatever you do is secondary to who you're becoming. God has given us an assignment and the assignment is this — we're going to the top, we can't be stopped, straight to the top! God has called us to be the best we can be!"

What I like about it is watching Caroll absorb it with true openness. I believe in little, except, strangely, I do believe in believers. I can't groove on the implied politics and Christian fundamentalism, but I can try to groove on the *groove*. Caroll is one of Pastor Keith's remaining original members, almost seven years after that first service in the school cafeteria. Church gives her hope, gives her strength and a sense of belonging. It gives her more friends. She arrives early and drops Marissa off in the "Surge" room (for kids up to fourth grade; at fifth grade you go to "Shock"), where Marissa is also a servant-leader of sorts, helping to baby-sit the infants and toddlers in "Wigglesville" and "Trailblazers." (Marissa loves babies, maybe more than she loves Coach handbags. Marissa cannot wait until her big sister has a baby in June.) Caroll then comes into the worship center, adjusts and tests her headset microphone, and takes her servant-leader spot in the center of the second row.

The lights go down, the drummer counts off "One! Two! Three! Four!" and suddenly Caroll is on her feet with everyone else, singing, clapping, bobbing up and down in place, singing this morning's Chrock anthem. Multimedia presentation is sacramental here. The lighting schemes are like the set of television's *Deal or No Deal*. When the music is finished, a "CCC News" program is shown on the video screens, packaged like CNN, with segments that perfectly mimic popular commercials and movie trailers, a copyright attorney's cease-and-desist dream waiting to happen. The worship stage gets a total makeover every few weeks, sometimes involving elaborate set pieces. Sermon series get titles, with themes and logos often derived from current hit TV shows (*24*) and movies (*Fantastic Four, Casino Royale*); Pastor Keith always casts himself as a fast-quipping badass warrior for Christ. He is not above driving a bulldozer on stage to make his point, which he recently did. He also likes telling the success stories of those who walk in Christ with him.

"There's a lady I know," he says one Sunday (there is always a man or woman he knows). "And she told me once, she said, 'God, will you use me to create something that could change the world for women?' She didn't know what, but she knew it was coming. She went away . . . and created a *purse light*. Some of you have bought it. It's a light and when you open your purse, it lights your whole purse. She took one word and she's made millions. Mary Kay loved her. There's things that come forth from here [Celebration Covenant], and let me tell you it is not about this building, it is not about this vessel, so please separate it from that, just let God work through you and watch what God might do in your life. Because the truth is, if you're going to think noble, you have to think opportunistically."

For her part, Pastor Sheila has sipped repeatedly from the cups of Oprah and Dr. Phil, while eschewing the garish mascara streaks and other leftovers from the Tammy Faye Bakker era. Pastor Sheila's specialty is connecting with women in the audience as divas and princesses. She stages a "Sister Sister" scripture and motivational workshop series for women, one of which was titled "If the Shoe Fits." She tells lots of jokes about shopping and how important it is to set goals as a mother, career woman, and warrior. At one "Sister Sister" teaching, Pastor Sheila reminds her flockette that the devil loves an airhead who isn't setting her life right with God: "If you're not working a plan and you're not planning your life, the enemy is going to set you up and he is going to steal, and he is going to kill, and he is going to destroy!"

Fire and brimstone are not, however, the Crafts' main currency; success is. "Attitude is the hinge the door of your destiny swings on," Pastor Keith says. That's just one of the "Leadershipologies" he claims authorship of. He will e-mail you one Leadershipology phrase per day when you sign up for more seminars.

At that first Sunday service I attend with Caroll, there's much excitement about the upcoming church Christmas pageant. Pastor Sheila calls Gina Anderson to the stage.

Gina, who is wearing a black turtleneck (message: she's artsy), is going to direct the pageant, drawing on what we are told is her ample experience in community theater. Gina says another church member came up with an original story for the script, and others are choosing the music. The entire stage, Gina says, will be transformed into a Victorian England town square. Gina needs carolers, children, actors, a snowman, dancers, and all sorts of technical assistance, and she needs it fast — they have only four weeks to get it together. There will be big sign-up lists after the service. (Caroll has already signed up to help with the technical crew. Marissa has signed up to play one of the extra children on stage.) Next comes the weekly money collection, at which church members are asked to tithe their "first fruits," or 10 percent of their gross income, which Caroll faithfully does.

Pastor Keith takes the stage. He is always in the middle of

a life-changing sermon series that will span many weeks or
months. Today could be part four of twelve (unless it is part
seven of ten, or part one of five). The series is often about "Xcel-
erating" to greatness, or "Mastering the Keys" (there are lots
of Keys — such as "Supernatural Keys to a Healthy Emotional
Quotient"). Or it's about the concepts of "Breakthrough," or
"Transformation." Elevations are big ("elevate your thinking")
as are eliminations ("negativity"). If you miss the powerful message one weekend, you can always download the podcast, or
wait until the end and buy the complete series on disc. In each
series, there are keys to the steps. There are steps within steps.
There are five ways. There are eight steps. There are four keys.

Very often in the many, many weeks I'll go to church here, I
will look over at Caroll, who is sometimes taking notes and nod-
ding along as if it all makes great sense. Pastor Keith will talk
for twenty minutes and then move on to point number 2, when
I'd assumed we must be to point 4 or 5 by now. My notes are often just sketches of what Pastor Sheila is wearing today. I am
lost.

Lost, but not unwelcome, for nothing pleases a church more
than a seeker. Pastor Keith encourages his flock to raise hands
in prayer and they do, but he sees one or two of us who are not
doing it. He is adamant, looking straight at us: "If someone
came up to you with a gun and told you to put your hands up,
you wouldn't say, 'Sorry, I just don't *do* that. It makes me feel
weird.' So please, put your hands up with us."

What the hell, then.

I lift my arms.

I hold out my palms.

I close my eyes.

The Gap

(A SLIDE SHOW)

L ET THE RECORD REFLECT that I did not come coldly to this Christmas present, bereft of God or lacking a Christmas past. I already had my own infancy narrative, my way-back-when, which was also set in an American elsewhere. I have the photographs, the brown lawns of my suburban childhood winters, the stray pieces of tinsel caught in the leafless trees, the houses just built on the new edge of the sprawling city.

I had the joy, too. I know the verses to all the songs. I know the Rankin/Bass stop-motion animation comforts of *Rudolph the Red-Nosed Reindeer* and *Frosty the Snowman*. I had great gulps of incense smoke wafting from the brass censer I carried as an altar boy at midnight Mass on Christmas Eve. I loved it all as much as the next kid, as much as you loved it, but perhaps not more, which could be where my unwrapping of all this started. My family shopped at the last minute, caught up in mobs, at malls that don't exist anymore — space-age shopping centers that faded from glory or found second lives with Vietnamese nail salons and *Quinceañera* dress shops. We sat in our cars and waited for an hour to get into the neighborhood by the lake, where the homeowners competed to put up the best light displays. There was that Dan Fogelberg song on the radio while we sat there, the Christmas tune about the guy in the grocery store — turn it up, please (*Met my old lover in the grocery store / The snow was falling Christmas Eve . . .*). Later, when I was in high school, there was a song on the radio by the Waitresses, the New Wave one about spending Christmas alone (*When what to my wondering eyes should appear . . . is that guy I've been chasing all year!*). That final Christmas in the old house, before we sold it, before my family started having Christmases in different places, mostly apart, there was Wham! on the radio (*Last Christmas, I gave you my heart / But the very next day, you gave it away . . .*).

The Christmases I grew up with (and away from) looked a lot like yours, regardless of your race or creed or income, because sooner or later, everyone's Christmas is put into the great pop-cultural compactor machine that Mannheim Steamrollers all holiday imagery and thoughts into the same flat shape and size. Despite magnanimous efforts to broaden the season to include other religions and cultures, the Christmas stories we tell ourselves tend toward white and suburban default settings. It's a series of scenes set to ad jingles that replace the lyrics of original carols with the brand names of what we all wish for. We guard our Christmas memories as unique and beautiful, damaged slightly in the shipment of memory, but even with a thousand tiny differences, these details are largely all about the same Christmas. Out of this emerge the clichés: the bad Tim Allen movies, the rote attendance taking, the listing of objects and traditions, our insistence that *we always do it this way*. People now travel to the Cleveland, Ohio, house where *A Christmas Story*, a 1983 movie about a Christmas in the 1940s, was filmed. That house symbolizes something now, and it is given sacred treatment each December 25 in twenty-four-hour rotation on cable television. To see these pilgrims, with their Ralphie Parker fixations, is to understand the way a religion catches on. A lot of times we honor Christmas most by doing it the way it is done on television, or as close as we can get to how it is done on television. And yet we insist our Christmas memories are sui generis. Every December we compare notes. We go over it and over it, like war veterans who never saw much action.

You say: *Here is what happened to my family one Christmas.*

I say: *Here is what my family always does.*

You say: *Here is a story about the year we . . .*

I say: *Here is what happened to us one year.*

You say: *Here is what we always did.*

Here is what happened, for a random example, during the Christmas of 1977, when I was in fourth grade.

Our family cat, a calico named Sociable, drowned in the almost-empty swimming pool in the backyard of a house up the street, in a foot or two of freezing water. The neighbors found her. My father went and got her and buried her the day after Christmas in the vacant acres behind our subdivision, and then I was told. I cried all day, and then the next day, and then the day after that. I tried to play with my new stuff — the Bionic Woman sports car with the "sabotaged brakes" action, so the Bionic Woman doll could stop the speeding car by sticking her platform-sandaled feet out on the concrete — but mostly I was wrapped in a child's sense of melodramatic grief. My eyes got crusty from crying. I played "A Star Is Born (Evergreen)" on the piano again and again. On the fourth day my mother told me that it was time to stop making such a show of myself. And she was right. Even now, the only thing I remember about Sociable is that she drowned in the neighbor's pool, on Christmas, and that I had overcried.

* * *

You say: *Tell me about Christmas when you were growing up.*

I say: *Dead cat.*

Oh, sweet and wildly inappropriate melancholy. There's this tendency I have to sit out the main event and carefully observe. Observing is a safe spot. Observing is sort of sad. A lot of times I can just call it journalism. Then comes Christmas, every year, when I again find myself standing outside a wide circle of fun that seems to include everyone else. In Frisco, I find that I can imagine some alternate life I never had: a wife, children, a huge house with a backyard with a gourmet barbecue grill. I've spent the better part of two decades accompanying and taking notes on everyday Americans engaged in their leisurely, passionate pursuits—and gotten nowhere with my own. I find that I relate no more to their spiritualism than I do to their love for pro sports. Why are they going to Las Vegas so much? Why do their weddings cost so much? Why flip-flops, why Crocs? Why bachelor parties, why tailgating parties, why *Girls Gone Wild*, why eat sticky cinnamon rolls in airports? Why all the righteous marathoning, walkathoning, triathloning? What are people talking about on all those tiny phones, all the time, and to whom? When did everyone get so busy? Why *The Da Vinci Code*? Why Carrie Underwood? Why George Bush? (Why Hillary Clinton?) Why paint your chest for a football game? Who are the people moving into those half-million-dollar studio condos? (Why don't they need a down payment? Who approved the loan?) Why plug-in scented candles?

I thought and wrote about these things, these innocuous yet fascinating matters of taste (plastic patio chairs, hair removal, beauty pageants), until I found myself slowly falling out of love with not only the elite side of life but the mainstream, too, growing more grouchy with America in all forms. I can be browsing in the Gap — any Gap in any mall, watching the college kids fold the sweaters, fold the jeans, fold the polo shirts — and be almost brought to tears by the perky music serenading me to the edge of the very gaping Gap-ness of it all. This gap has grown to a chasm.

There is obviously some problem here, a pose I'm too good at projecting, some predictable dissonance of cynicism. Something about Christmas is packed away in a box so deep in the garage of my psyche that I hardly know where to find it. I think about these things a lot now, crisscrossing Frisco's strip mall parking lots every day, going to church with Caroll, decorating trees with Tammie, watching the lights at Jeff and Bridgette's house. This constant observing, this vicarious longing — for what?

Choruses of angels are not harking and heralding for me. I prefer dark, slightly twisted Christmases. I like the doleful notes in the Vince Guaraldi jazz score of *A Charlie Brown Christmas*. Sleigh bells ring, and I sigh heavily for no reason. People like me (and there are many of us, all thinking we're uniquely alone, which makes us still another niche in holiday marketing) believe in sour candy. We believe that David Sedaris in his department-store elf outfit, Billy Bob Thornton in *Bad Santa*, and the creators of other misanthropic, antiseasonal fare are really *on to something*. Someone had to start the urban myth about suicide rates increasing at the end of the year, but who did? American suicide rates actually peak in spring and fall, according to the National Center for Health Statistics, which every year has to debunk, via press release, the myth of Yuleicide. December, in fact, has the *lowest* rate of suicide. Some still want to believe it and report it as fact, for what better time to check out than Christmas? Why the blues, in all that twinkling light?

Science says maybe it's something primal in us. We get so

cozy in the winter that we become sad. We hear things in the crystal-clear night. Branches snap, the fire flickers, and millennia later, here we are, under the chenille Pottery Barn blankets of what passes for deeper significance.

I grew up with bountiful, pleasingly middle-class Christmases in a large, extended family. A couple more slides from my Kodak Carousel projector of the personal:

Here is my mother, on evenings near Christmas, tuning in carols on the classical FM station on the oak console hi-fi. Here's an Advent wreath on the kitchen table, the candles flickering each night at our kitchen table — purple, purple, pink, purple, "O come, O come, Emmanuel" — another great song made up entirely of sad notes, about the desperate sense of longing in the Israelites.

My memories are carbed up with Chex Party Mix just out of the oven, spread on Bounty paper towels to cool and soak up the excess butter, one Crock-Pot filled with melted Velveeta and Ro-Tel diced tomatoes, another filled with cocktail weenies aswim in barbecue sauce. I happily awaited the holiday nights that our four-bedroom, two-and-a-half-bath Oklahoma City tract house reached maximum capacity when everyone came home, all my sisters and their husbands and new babies — when the sofa beds were pulled out, bedrooms reassigned, my sisters' L'eggs pantyhose hanging on the shower rod, their curling irons plugged in all day. There are hundreds of snapshots of the people I love, with sleepy heads and wide-framed seventies eyeglasses, holding up just-opened presents for the lens of my father's camera: yellow bathrobe, remote-controlled car, air rifle, roller skates, cowl-neck sweaters from J. C. Penney, a James Taylor LP. Even after my parents' divorce, there are still more pictures of people unwrapping presents and holding them up for the camera. We loved and we fought and we gave and we got, and we took it back to the stores if it wasn't what we wanted.

Twenty years ago, my siblings and parents and I all started heading our separate ways, remaining fond of one another via the Christmases of Verizon and UPS. I'd like to be able to point to some family cataclysm and say, *This is why we don't do Christmas together anymore,* but there's really no one reason. What happened was we became Santa Claus for other people.

After college, I started working as a newspaper reporter and frequently was assigned the Christmas Day shift on the city desk — making the cop and fire dispatch checks, listening to the scanner, going down to the multiplex one year to interview people who spend Christmas Day at matinees, or off to interview people stuck at the airport. Another year, it was the task of knocking on the door of the house where the relatives had been killed in a drunk-driving crash on Christmas Eve. Once, after one of those shifts, my big sister and I met at Bennigan's, next to the mall's empty parking lot, over glasses of white wine. That was the year she'd just split from her husband. It was the only restaurant open on December 25, and hardly anyone was there. By 1991, Christmas seemed to be happening to everyone else.

And so, like a family going in different directions toward their own tastes, I sensed I had drifted too far from the very culture I claimed to understand so deeply. America was hanging out in places I deliberately wasn't. It was going to churches the size of hub airport terminals, and to Céline Dion and Toby Keith concerts; to Hooters, to Abercrombie & Fitch, to Dave & Buster's; it was congregating in Bed Bath & Beyond parking lots on Monday nights to show off its customized trucks and pimped-out Chevy Cavaliers. America was going all out for Christmas every year, and why wasn't I?

Here is one last photograph, in the very last shoebox I searched. It is December 1974. A pallid department-store Santa in a crooked beard, wearing a cheap, loose-looking wristwatch on his left wrist. His right arm is squeezed around a six-year-old, blond-banged boy in a red coat and blue striped sweater. I look at this picture now and see a boy doing everything he can to play along.

Manger Babies,
Angel Trees & Tiny Tims

✦

(ONLY 23

SHOPPING DAYS

TILL CHRISTMAS!)

8

Mary, Did You Know?

TAMMIE CALLS. "Elf!" she says. "Where are you? Get *over* here and see what I'm doing. I am in such a zone today!"

Between my frequent trips to Stonebriar Centre and church services, and whenever I'm not investigating the progress of live Nativity *posadas* marches or neighborhood cookie exchanges, and while I'm not tracking stock prices on major retail corporations or needlessly noting the pre-holiday markdowns at Macy's, I've been helping Tammie Parnell decorate interiors. She is averaging at least one house a day, including weekends.

When Tammie calls and insists I stop what I'm doing and come to whatever house she's currently working in, I go. Sometimes I'm there to assist. Sometimes it's because she needs company. Sometimes I am summoned to marvel at her handiwork. Tammie gives me general directions on the phone as I drive: "Past the strip center with the Mi Cocina restaurant, past that new BMW dealership," she will declare, thinking that is enough, and then rattle off a few more details. "Oh, gosh, here's what you do. Get on the tollway. No, don't —" Take Headquarters Drive, she says. No, wait, Spring Creek Parkway to Shadow Rock, Shady Point, Whispering Woods. (The street names start to sound as if I'm following Nancy Drew.) "Just drive until you see my car." Winding, winding, into a higher-end neighborhood of new 5,000-square-footers and carriage-style driveways with three-car garages.

With one or two large-scale exceptions, the houses Tammie decorates could all be the same house. Architecturally, the Dallas suburbs seem oddly determined to become a replica of the English countryside, a posh world of make-believe polo matches and fox hunts. The houses all have the fieldstone veneers and the eight-foot mahogany-stained doors, the rounded interior Sheetrock corners, the skip-trowel wall finishes, the wrought-iron balusters on the foyer stair. They have the custom ash or oak cabinetry, the coffered ceilings in the master, the double-stack crown molding, and the great-room barn rafters of allegedly reclaimed wood. They are the luxury homebuilding equivalent of the boob job. Tammie is teaching me to see hints of the unique where I previously saw only lockstep conformity and a craving to impress. It's that middle-of-America sense of wealth and taste, hatched somewhere between old-money Dallas and the forever stretch of prairie. The sofas and chairs are plump. The end tables, sideboards, and dining tables look handcrafted and heavy, as if they arrived on some tall ship from an imagined Old World.

The aesthetic is at once "antique" and "Tuscan" and "Ivy League," and the color tones are all HGTV correct: dark woods with autumnal and leather hues in the family room; creamy, sunny tints in the living room and dining room; fairway green or burgundy in the husband's study, with framed watercolors of notorious holes at famous golf courses; lots of princess pink pastels in the girl's bedroom; a sporty or cowboy motif in tan, blue, and red for the boy; stainless-steel appliances and granite countertops in the kitchen, with cabinetry reaching all the way up to the tray ceiling; finally, the hand-scraped oak floorboards underfoot. The art on the walls fits right in with Christmas — expensive photo portraits of the family here; a Thomas Kinkade cottage there. It could be Tammie's house. "This blanket," Tammie says of a sumptuously thick throw resting on a leather chair across from a Christmas tree we are decorating in one of her clients' homes, "this is a $500 blanket. Horchow, maybe." This blanket evokes all the right things: curling up with the classics, while a pretend snow falls outside.

Tammie adores rich people. She loves horse country, Sunday mimosa brunches at the club, dress codes. She can't stand it, though, when she thinks rich people are being too selfish. One of her earliest jobs this year is in a house Tammie values at well past the million-dollar mark, which is occupied by a client Tammie is convinced doesn't care about anyone but herself. Tammie notices little things, such as how bossy the client is — "How ungrateful," she says, to complain that her tree looks crooked after Tammie has finished. (Tammie went back and straightened it.) Tammie used all sorts of decorations in several rooms of that house — pricey ornaments, greenery, vases, *Nutcracker* soldiers, ribbon, the works. But there was just no Christmas there.

"See, that's my thing with her, though," Tammie says. "I don't think they're believers in anything, which I sort of have an issue with. Believe in something. Whether you're Buddhist or whether you're Catholic. Just believe in something. Believe, you know, in being charitable. You are *blessed*. I was doing a house Sunday night, really beautiful, really nice people . . . [The husband] has done phenomenally, he's got a company that manages hotels. He's in Arizona right now, you know, playing golf in Scottsdale, but you know the first thing he said to me was when I got in this business, he says, 'Tammie, you realize that you're blessed and you share it with other people.' So true."

It's usually at this point in her grueling season that Tammie wonders if she hasn't shared too much of her own blessings with everyone else. "Another person called me up and said, 'Please come do my house,' and I had to tell her [no]. I have to stop eventually, with the houses I've already got lined up. I can't keep doing it, or my own family is suffering pretty soon."

Tammie is discreet in these unpeopled houses. ("Don't write this down," she will say on some jobs, launching into some personal story about the client.) In each house there is the potential for some other narrative thread. You start to know the kind of information the housekeepers know, with that strange feeling of handling other people's boxes marked "Christmas," moving their furniture around to accommodate a tree, looking at the photos displayed on their shelves, sorting through the detritus in their garages. In one house, Tammie sends me on a hunt for "any sort of tray" on which she can arrange a set of snowman glassware. There are no trays in the wet-bar cabinets, or in the TV room, or in the dining room buffet, so she sends me upstairs, where the family's two boys live in adolescent splendor: two large bedrooms flank a media room that has a large sectional sofa, a huge flat-screen TV, a computer, and several video game consoles. I don't find a tray there, but in the bathroom, in a lingering fog of Axe body spray, I notice a chrome trash pail next to the toilet stuffed high with used, teenage-boy-sized disposable diapers. I retreat back downstairs. "Find anything?" Tammie asks.

"No luck," I say.

When the clients come home in the early evening, Tammie has vanished, having wondrously left Christmas in her wake. There's a personalized, inspirational note from Tammie on the foyer table, with an old-fashioned brass skeleton key tied with red ribbon. The note is about discovering the key to the season's blessings.

"You'll love her," Tammie says of a different day's client. "She's really down-to-earth, really likable." Trish is a stay-at-home mom who has hired Tammie to set up and decorate her home's formal living room tree, put the garlands and trimmings on fireplace mantels in both the living room and family room, and decorate the dining room and the large entry hall. She'll also install the prelit garland all the way up the stairway rail.

Tammie has spent the first part of the morning with Trish, sorting through the family's Christmas decorations. She's talked Trish into buying a few basic but transformative items from Tammie's own supply — mostly ribbons, runners, and two angel figurines. Tammie has on her Two Elves with a Twist apron, which holds her scissors, pliers, and other tools, and she is laying out plastic garlands along the floor.

Tammie's most time-intensive service is garland-wrapped staircases, which usually take about two hours to complete. The garland often has tiny white light bulbs in it, and Tammie obsesses over how it is wrapped, which way the light bulbs point, and what sorts of floral wrap ribbon, glazed plastic fruit, coral cranberry beads, and other thematic touches should finish it off. The plastic of artificial garlands and Christmas trees is sharper and sometimes more prickly than actual pine needles. A day spent decorating scratches up Tammie's hands.

Trish is in her late thirties, with layered blond hair; she wears a lot of foundation. Her husband is an attorney. They have three children, two girls and a boy. "I just couldn't do it this year," Trish says — the decorating, the unpacking of Christmas stuff, the *thinking* that's involved. She couldn't face her same blah ornaments and knickknacks, putting them on the same tree the same way. "With no theme," she laments. "I needed help with that."

Trish has also been sick and tells Tammie and me about the rare, complicated bone marrow deficiency with which she's been diagnosed. "It's like cancer, but it's different," she says, letting it

119

hang there, sadly. In the last year it's meant a lot of doctors, blood tests, treatments.

Tammie clenches her jaw, squints, and nods sympathetically. "Well, I am here to make your Christmas better," she says, checking her watch (10:30). "Let's get to it."

Trish brightens then and shows us her "amazing find" from Bed Bath & Beyond: fourteen identical gift baskets with a "movie night" theme that were on sale for $21.99 each. The gift baskets consist of a ceramic version of a red-striped, concession-stand popcorn box that contains Raisinets, Junior Mints, and microwave popcorn. Trish plans to unwrap the cellophane on each basket and "Christmas them up" by including a DVD. Some of her gift basket recipients will get *It's a Wonderful Life*, the 1942 Frank Capra holiday classic, and some will get *Love, Actually*, the 2003 British ensemble comedy starring Hugh Grant, which Trish likes better, because every time she watches it, she cries. (*Love, Actually* seems to be the preferred, watch-it-over-and-over Christmas movie of every woman I meet during the Christmas of 2006, because it features several recognizable crises: marital woe, loneliness, family estrangement, and grief over the recent death of a young mother, all woven into romantic story lines. *Love, Actually* is drenched in that particular upper-class English coziness that brought Christmas to America in the first place; it has gorgeous furniture, cute clothes, happy yet melancholy endings, British accents, lots of amber light in living rooms, and a well-timed snowfall.)

"That is an absolutely phenomenal idea," Tammie coos. "Movie gift baskets! What a find! You did good, girl."

Trish is pleased. I get ready to sort through two tubs of ornaments, separating them (as Tammie has requested) by color and setting aside most of the "sentimental" one-offs that will not make Tammie's edit on the living room tree. Trish soon retreats to the master bedroom and shuts the door, to leave us to our work.

After Tammie is done testing the bulbs on the strings of stairway garlands, she turns to the mantel and the nine-foot tree in the formal living room. She has assembled it and fluffed its plastic branches, stood it up straight, plugged it in, and prepped it for the best trimming of its manufactured life.

With the gold ornaments I've sorted, she and I decorate the tree for almost an hour. In addition to ribbon striped in autumn hues, and pieces of gold-flecked floral wrap that she scrunches among branches with a calculated yet random effect, Tammie uses lots of feathers, along with expensive ornaments in the shapes of birds. Her favorite trick on any tree is to use several long pieces of "ting-ting," which is golden wire twisted into a spring-like curlicue. Ting-ting is Tammie's tinsel — and so much better-looking, she says. She has me reach to the top and stick in the uppermost pieces of ting-ting. The tree is now bursting with charm.

Next comes the mantel. Tammie has brought along two of her favorite finds this year — eighteen-inch-tall angels in elegant burgundy satin robes, with their wings majestically outstretched. "Aren't these phenomenal?" she asks. "Hobby Lobby. Over on 15th and Alma. I walked in there and I just had to have them. I wish they had more. I have at least five clients I could have sold these to."

Tammie and I twist and arrange the mantel garland. She whips out two rolls of ribbon and wants to know which color and pattern I think she should scrunch into it, and how much of it, and which way. I stare at both choices with a true attempt at discernment: More gold? Or the rust-and-gold plaid? I tell her I'm no expert. If I were an exurban housewife married to a golf nut, with three kids and a problem with my blood-platelet count, I would have long since hired a Tammie type to rescue me. I'd hire Mexicans to hang the lights outside. I'd hire a caterer for my neighborhood party. I'd have my family portrait for the Christmas cards made in September, with us wearing our matching rugby shirts, print out the addresses on label-maker software, and ask my teenage babysitter to stuff the envelopes for an extra twenty bucks. I'd pop *Love, Actually* into the DVD player every day for a month and medicate my way through Christmas with eyes wide shut. *Fake is okay here.*

"Tammie, I don't know," I say at last, in a burst of opinion I never knew I held. "I'm thinking you should use the one that's more gold, and this angel should be farther to the right, not in the center. I think the other angel should be over here, on the table, in a centerpiece. That way the whole room starts working together."

Tammie is silent. She squints at the mantel, then her eyes dart to the table, then the tree, then the mantel, then the angels, then the mantel again.

"You're really starting to understand your garlands!" she suddenly squeaks. "I need you. I need you to say what you think. You've got the eye, mister."

Refreshingly, Tammie harbors no pretense that her work is a mission of mercy for the Christmas-impaired. Getting into the true spirit is up to the clients; all she can do is make the chores more manageable. "This is a *business,*" she says, again and again. "I'm running a business." Tammie earned her master's degree at the Thunderbird School of Global Management outside Phoenix and her undergraduate degree at Clemson University. She prizes her own marketing savvy almost as much as her role as a mother, wife, and doer of God's work. What seems to her now like a lifetime ago, Tammie was a menswear buyer for

Federated Department Stores, based in Miami. It was the eighties. Tammie pushed the *Miami Vice* look on American shoppers, with lots of Creamsicle pastels. She once traveled the country giving presentations on nubby, unconstructed tweeds for one season and on paisley vests for the next. She always kept an overnight bag packed and ready to go in the trunk of her little convertible. She felt completely free. Two Elves with a Twist exists mainly because Tammie longs to once again seize corporate America by the lapels. "Someone's got it," Tammie says, recalling the voice of one of her early mentors in business. "And if you see a dollar that can be made and it's not being made, you've gotta grab it."

Tammie will insist all afternoon that she is "almost done" with Trish's house, but it will be long past dark before she actually finishes. Her "time is money" principle often wavers. She is always answering her cell phone — other clients are calling, or Tammie's mother is calling from Florida, or Blake forgot something for school — and as she does, she'll stop one task (arranging snowman cookie jars) and pick up another task (rearranging a centerpiece on the dining table). She could hire another elf, "but hiring someone to help me takes away from my bottom line," she says. "If I'm averaging $100 an hour, and I have to pay someone else, and factor in the supplies, even with my markup, pretty soon it eats up my profits."

The original estimate Tammie wrote up for Trish's house was for six hours and, with materials, came to $700. But those six hours stretched to ten. Tammie kept adding ribbons and other objects, each with Trish's okay. With the extra hours, the final bill came closer to $1,000. By nightfall, the house had taken on the honey-hued glow of Christmas — just like something out of *Love, Actually*. (Clients are always paying Tammie the same compliment: *Our house looks like a page in a magazine.*) Tammie tells me later that Trish went to her bedroom, returned with a wad of cash, and asked if she could write Tammie a check for the original estimate amount and make up the difference in cash. "She said, 'I gotta do this under the radar,' and I said, 'Oh, honey, I am all about under the radar!'" (Couples are always hiding purchases from each other, Tammie notes: new golf cleats in the trunk of his car, new clothes sneaking into her closet under dry-cleaning plastic bags.)

Whether their budget radars are reading right, it's the husbands who are sometimes Tammie's biggest fans. One client's husband called Tammie last week and raved for ten minutes about how, *for once*, his house looked like the perfect Christmas and his wife was not a perfect wreck.

Other husbands never notice the decorations. For all his crisp style and attention to details, Tammie's husband, Tad, may be one of those husbands who are generally oblivious to the allure of ting-ting or the differences between better-made plastic garlands and the stuff at Wal-Mart. If Christmas is indeed the most important time for family, Tad reasons, why does Tammie spend so much time in *other* people's houses, making their Christmases easier, but making her own family's more chaotic? Tammie's energy level is something that made Tad fall in love with her, back when they were both in the garment trade. But when is it all too much? By the second week of Tammie's work, she and Tad have usually had an argument or two.

Tammie gets home long after Emily and Blake have gone to bed. "Tad understands it, and he doesn't," Tammie says. She spends another half-hour or so in the garage sorting garlands and other materials into tubs for the next day's job.

Dear Santa, I can explain: "Every relationship has a show dog," she says. "In every house I'm in, or whenever I meet a couple, I try to figure out which one is the show dog." In her own marriage, Tammie is the show dog. She's always moving, talking, getting into people's lives, and finding too much to do.

Christmas is her show-doggiest time of year, and she just can't help it. I ask her again and again what the allure is, putting pieces of mass-manufactured plastic up in other people's houses. Sometimes, when Tammie is less guarded, the answer comes back around to the extra cash. When she sees me taking notes, the answer is always how much she just loves Christmas.

After midnight, Tammie soaks in the Jacuzzi, with a glass of Chardonnay. She never looks more than a day ahead on the calendar in her decorating appointment book. "Otherwise I get too upset about how many I have left to do," she says. (I look. There are twenty-six houses left to do in the next twenty or so days. Tomorrow is Candy's house, which excites Tammie, because earlier in the day, while at Trish's, Tammie had called Candy and talked her into paying for all-new greenery. "Girl, we have got to replace your garlands," she'd implored.)

Just before crawling into bed, Tammie slathers her nicked-up hands in a thick layer of a lotion she likes from Bath and Body Works. She puts a sock over each hand and asks Tad to get the light. She sleeps three hours, sometimes four.

The days are bright and warm, a perplexing December spring. Tammie calls and again tells me to *get over here:* Preston Road to Park Boulevard to Willow Bend to Bridal Bend, winding, backtracking, cul-de-sacking, looking for Big Red parked out front, lured here by Tammie's imploring, *You've got to see what I'm doing in this house.*

The houses are like another glossy page out of *Plano Profile* magazine. "She really loves animal prints," Tammie says in one, by way of explaining a tree with leopard-spotted ribbons and more feathers.

"She loves it when I hang these [Christopher] Radko ornaments from the chandelier," Tammie says in another.

"She's all about these *snow people*," Tammie says in yet another house, with a loathsome regard for a group of four-foot-tall snowmen and snowwomen carolers. She'd just as soon banish them to the garage but instead compliantly arranges them near the stairs, just the way the client likes.

To set the mood at most of her jobs, Tammie turns the radio on. Usually, there is a stereo system piped through the entire house. She finds KLTY, 94.9 on the FM dial, which bills itself as the nation's largest Christian radio station (motto: "Safe for the Whole Family"). KLTY plays Christmas music twenty-four hours a day, starting in mid-November. Between songs, KLTY runs little messages from its listeners thanking the station for "keeping the Christ in Christmas," the reason for the season, and such. It runs a station identification between songs with the emphatic voice of a little girl reading the Nativity passage from Luke 2:11 — *For unto you is born this day in the city of David a Saviour, which is Christ the Lord* — and she really hits it hard with *Cah-rice-stuh the Lo-ward-uh*. The sound of a child reading the Gospel in a southern twang pleases KLTY listeners no end, their own Tokyo Rose in the war against the war on Christmas.

Every time KLTY is on, I hear one or another version of "Mary, Did You Know?" — a Christian radio hit first released in 1991 and then remade by everyone from Reba McEntire to Donny Osmond to Kenny Rogers to Natalie Cole to Clay Aiken and others. The first forty times I heard "Mary, Did You Know?" it made me itchy. Now I surrender and hum along:

> *Mary, did you know*
> *That your baby boy*
> *Will one day walk on water?*
> *Mary, did you know*
> *That your baby boy*
> *Will save our sons and daughters?*

Did you know
That your baby boy
Has come to make you new?
This child that you've delivered
Will soon deliver you.

Even with the Christmas music on, Tammie gets lonely in these monster houses. She talks to the elves and snowmen and Santas. She calls them "fella" and "mister" and "big boys" and "bad boys." "You, mister, are going right here," she'll say to a snowman on a console table in the entry hall, and then pick up a porcelain Santa. "Okay, big boy, where are you going?" She carries on entire conversations with the pieces of a Nativity scene. "You are such a good mother," she'll tell the Virgin Mary. She talks to angels, camels, and donkeys. She has a special relationship with Balthasar, Caspar, and Melchior — the three Wise Men — and calls them each "sir." Tammie has conversations with the Wise Men about which side of the dining room buffet they want to stand on — "Yes, sir, you belong right here, what do you have to say about that?" — and then wonders if one of them should be standing higher than the other two, on a book propped up under a cloth.

Books are a useful resource in a Christmas decoration job. Tammie will go through a client's house and scoop up hardcover books for decorating purposes, especially to assemble a more suitable crèche setting, with books stacked under perfectly wrinkled table linens or spare runners, fashioning hills of Bethlehem. But books are sometimes hard to find. There are plenty of nice built-in shelves in each house she decorates, but the owners tend to fill them with vases, family photos in expensive frames, and fake plants. If we luck out, there's a series of gold-leaf leather-bound classics or an *Encyclopaedia Britannica* set, or perhaps the husband kept his law school books. There is always a nice, big Bible, but Tammie won't stack on a Bible. Some houses possess quite a few hardcover airport thrillers that work okay under a cloth. On some jobs, Tammie sends me on a mission throughout the house to find any book thicker than two inches, and I'll return with a stack of insufficiently slim epistles by business gurus and self-help spiritual gazillionaires such as Zig Ziglar or Joel Osteen. In a music conservatory occupied by a dusty piano in one house, I press on a concealed closet door, and when it opens, a light automatically clicks on and I discover a row of hardcover Christian adventure stories about the Rapture. "Perfect," Tammie says.

For both of us, all work stops whenever a KLTY "Christmas Wish" comes on the radio. Tammie got me hooked on them. I'll be putting ornaments on the top of a tree while Tammie's in the kitchen arranging Santa Claus coffee mugs on the buffet counter, and we both stop and listen to the Christmas Wish segment, which airs a few times a day.

It works like this: Hundreds of listeners send e-mails, faxes, and letters to KLTY with the saddest, neediest case of someone they personally know. A Christmas Wish often involves a stoic child. Disease and a lack of health insurance are involved, mixed in with divorce. A car has broken down; someone has lost a job; the bank is about to foreclose on the mortgage. Christmas Wish seeks especially those people who are too proud to ask for help, underscoring the KLTY listener's disapproval of taxpayer-funded support. The station claims to have given away some $750,000 worth of donated money, goods, and services in 2006, and the recipients are always hard-working, faithful, and simply unlucky — thereby deemed deserving of the private sector's blessings and cash prizes.

A popular KLTY on-air host, Frank Reed, conducts three-way conference calls among himself, the wisher/sufferer, and the person or business chosen to grant the wish. Here's one such Christmas Wish, which KLTY must have considered a prime example of the genre, since it was posted as an audio clip on the station's website.

Frank reads it from the beginning:

My name is Angel, and I teach at a daycare, and one of my students is a little boy named Joshua . . . He is one of the most loving, sensitive, and giving little boys that I have ever known. He is the best big brother in the world to his little sister and he takes care of her all the time. When other children made fun of Josh for believing in Santa, [he] sweetly told them that there is a Santa, and he exists in the spirit of Jesus . . .

Angel's letter goes on to say that Joshua's mother fled his father, taking Joshua and his little sister with her. Frank continues reading Angel's letter: "[Joshua] told me . . . , 'My daddy hit me, and my mommy and me had to leave. We now live with Grandma and Grandpa and that's good, because now we are safe and nobody can hurt us again.'"

Here, Frank begins to cry, but you can also hear heavy breathing and sniffling as the person Frank has waiting on the other line listens to the letter, presumably for the first time. Here comes the disease part: "Recently, Joshua's mother was in the hospital and the news is not good. She's been diagnosed with a virus that could be fatal; she's been in and out of the hospital, and each time I see her, she looks weaker and worse than the last. She's got a new job, but I'm afraid she can't afford the insurance that would help her receive the necessary treatments she needs to stay alive . . ."

At this point, Frank dramatically builds it out for his listeners. The teary person on the other end of the line is Mandy —Mandy, who is Joshua's mother! Frank then tells us that

Joshua wrote a letter and asked Angel to "deliver it to Jesus," but instead, Angel very smartly sent it to KLTY.

"Mandy, are you sitting down?" Frank asks.

"Yes," comes a blubbering response.

Frank is now going to read Joshua's letter. *"Dear Jesus,"* it begins:

My mommy is sick again. She's sad because she says she can't give me and my sister the things that we want. Could you please tell Santa this year to bring my sissy and me some warm clothes? My mommy has no jacket — it's cold outside and she needs a new mattress for the bed. And if it's okay, could you bring us two bikes? Or, if that's too much, we could share a bike and I could give my mommy my bed and sleep with my sissy. I don't want my mommy to cry and if she cries she might get sick again and the medicine is real expensive. The doctor said my mommy might not get better so please help my mommy be happy so maybe she can get better. I love you, Jesus, and I know you love me too . . .

"You have been pouring some incredible stuff into that young man," Frank tells Mandy. "He is something else."

MANDY: He is.

FRANK REED: Well, Mandy, obviously the needs of this Christmas Wish are very, very large, but we have hooked up with some wonderful people. On the telephone, I'd like you to meet the senior pastor of Highland Meadows Church in Colleyville, Drew Sherman.

DREW SHERMAN: Hi, Mandy.

FRANK REED: Drew, since I am losing it, I'm just going to give over the radio [airwaves] to you.

DREW SHERMAN: I understand — thanks, Frank. Well, Mandy . . . we know [your insurance] is very expensive, and

we've heard that to cover you and the children it'd be
about $400 a month, is that right?

MANDY [sniffling]: Yes.

DREW SHERMAN: Well, we just want you to know, first of all,
we just love ya — and we don't know ya — but we love the
Lord, and out of the kindness of the Lord Jesus, we want
you to know we're going to pay that insurance for a year
for ya, because we want to make sure you stay healthy.
Also we're going to make sure that Joshua gets that bike,
and his sister, and we're going to be taking care of [the
gifts for] the children as well . . .

The pastor then says that the church is also going to get
Mandy a new mattress! And then he wants to know what else
she needs for herself and the kids. Do they need food? "You need
to let us know," he says. "We just want you to know that we love
ya, and we're going to take care of ya this Christmas."

MANDY: Oh, thank you so much!

DREW SHERMAN: I wish I could take credit for it, but this is
the body of Christ at work here at Highland Meadows, so
you'll be a real blessing to us.

FRANK REED [gathering himself]: Mandy, this is a wonderful group at Highland Meadows. And also, Mandy, you
can know that [there] are thousands and thousands of
people listening to 94.9 KLTY here, and they are gonna
be praying for your health, they're gonna pray that you
recover completely, they're gonna pray that you get back
on your feet, and we just want to say Merry Christmas
from Christmas Wish.

MANDY: Oh, thank you. Merry Christmas and bless you!

FRANK REED: Everybody just hold on. We all need to take a
big, deep breath . . .

* * *

There is at Christmastime a powerful currency in the anecdotal. The pastors in the churches, the talk-show hosts, and the motivational speakers seem especially adroit at encountering the suffering of first-name-only (or anonymous) charity cases who are always presciently good at telling their humble narratives of Christmas woe. They always say just the right thing, as if written into life by Charles Dickens himself.

The KLTY Christmas Wish kills Tammie and me every time it comes on, but in markedly different ways: Tammie hears it, carefully wipes a tear away with the edge of her index finger, and is immediately moved to find a fellow human being whom she can bless with a Target gift card. I hear a KLTY Christmas Wish and have all these questions — questions I don't exactly require answers to so much as I need for them to be asked. This disqualifies me from membership in the congregation of conspicuous Christian love, because such questions are the act of a skeptic, and in this realm a skeptic is no better than the Grinch.

Nevertheless, why wasn't Joshua himself on the air? This articulate "little man of God" sounds like he'd be great on the radio. Have any corners been cut in this story, such as Mandy's work history, restraining orders filed on the abusive husband, and all that? What is the medical term for the viral condition from which Mandy suffers? Are her insurance costs incurred through a policy deductible offered at her new place of employment, or through an independent provider? The grandparents with whom Mandy and her children are living — are they also struggling financially? Did Joshua really tell the schoolkids about how Jesus and Santa work in tandem, or did Angel, the teacher who says she has known Joshua only since July, embellish this part? Where in the Dallas/Fort Worth metropolitan area is this saga unfolding? (Note to self: This is why people hate the godless news media!)

Beginning in the mid-nineteenth century, British and American novelists excelled at illuminating the meaning of Christmas through saintly fictional children, who were supposed to inspire the same in actual children and their parents. Victorian sensibilities abhorred the selfish child, and since then, our culture has similarly treasured the little one who wisely puts the Christmas wishes of others ahead of his or her own. Dickens gave us Tiny Tim and the other Cratchits in *A Christmas Carol* in 1843, and they remain the pinnacle of humble, seasonal accord. Louisa May Alcott gave us *Little Women* in 1868 — the March daughters, who, rather than focusing on their own Christmastime joy and sorrows, are compelled to give to others.

Practitioners of nonfiction, later to be known as "the media," have similarly worked overtime in December to provide readers and viewers with proof that such children exist — cherubim among us who care not for Nintendo DS so much as they care for their fellow waifs. We crave that child at Christmastime. In the late 1800s, according to historian Stephen Nissenbaum's *The Battle for Christmas,* rich New Yorkers used to buy tickets to watch poor children receive Christmas presents and eat donated food at banquet spreads in Madison Square Garden. Nissenbaum cites newspaper accounts of such grand shows of charity, noting that some of these events were designed to get rich children involved with acts of kindness to poor children, but the rich children weren't as interested. It was the grownups who ached to connect to the poor children, to witness a Yuletide miracle. Almost as soon as we invented our Christmas culture, we acquired a hole in our hearts that only wistful TV specials can fill. KLTY smartly matches its advertisers with wishes that they can then, in essence, buy the opportunity to grant to these modern equivalents of the ragamuffin.

I wish I could say I spend the rest of the afternoon thinking about Mandy and Joshua, but I don't. "Mary, Did You Know?" comes on yet again, just another afternoon in another massive house with no one in it but Tammie, me, and a geriatric schnauzer. I hum along, realizing that the Santas and Wise Men and holly stuff in each of these houses will always look sterile, meaningless to me. Tammie is seeing something I'm not.

What I mean, Mary, is that I *don't* know. It is entirely possible that I suck at all of this Christmasy goodness. This is what I'm thinking to myself, even as Tammie and I have a long and earnest discussion about where to stick a big angel.

9

Restoration Hardware

WITH JEFF GONE all the time, overseeing the installation and programming of the Christmas lights at Frisco Square, Bridgette Trykoski focuses her attention on her own décor. She takes a day off work in late November, turns on *Oprah*, and unpacks the Christmas boxes from the spare closet in the guest bedroom. Oprah has Dr. Ahmet Oz on again, talking about the shape and size of normal, healthy excrement. "Mine are shaped like *C*s," Oprah declares, when Dr. Oz states that perfect bowel movements should be shaped like *S*s. "Any letter is good," the doctor says. "You just don't want apostrophes." Oprah encourages her minions to be sure to look at their own.

Bridgette and Jeff went to Lowe's and bought an eight-foot tree for $40. It is a real tree, the only real tree in a home I've encountered so far this year, even though there's a tree lot at every other intersection. (In the American living room, the number of fake trees first surpassed the number of real trees in 1991. More than 70 million families own fake trees, which have gotten fancier, taller, fluffier, and more expensive. In 2006, 28.6 million real Christmas trees are sold in the United States, for about $1.2 billion, down from a recent peak of almost 33 million trees the year before. Between 9 million and 10 million fake trees were sold this same year, and they will last an average of six Christmases before they are replaced.)

"Would you ever guess that I put Disney princesses all over my tree?" Bridgette asks, knowing by this point I would. When the Hallmark ornament catalog comes out in July, Bridgette carefully pages through it and circles the ones she wants, as well as ornaments she'll give to relatives and friends. She buys nearly every Disney princess Hallmark issues.

Jeff's lights projects get busiest just when the e-mails start going around among their families about this year's Christmas gifts. The Trykoski side of the family draws names in "Secret Santa" style, with a lot of back-and-forth among participants online about what to not-so-secretly buy from one another's not-terribly-secret wish lists.

In addition to that, Bridgette herself gets lots of Christmas presents from Jeff and from her parents — the joke on her side of the family is that Bridgette's pile is still the largest, even this year, with her little nephew sure to get a lot of toys. There's a fifty-fifty chance that the gifts Bridgette receives will need to be returned, especially what Jeff buys for her. ("Remember the Banana Republic disaster?" he asks, which she both confirms and ignores with another deft rolling of her eyes.)

Bridgette has an array of little emotional buttons, each blinking and ready to be pressed at the presser's peril. Her husband enjoys the risk. To an outsider, it seems she bosses Jeff around a lot. She doesn't drop hints for presents so much as make specific requests for items from the Coach handbag store and Jared jewelers. ("Is there still that one salesgirl at Jared with the great tits?" Greg wonders aloud, once, interrupting Bridgette's litany of what she's asking for this year. "Shut up, I'm talking," she tells him.)

Eventually, the princess saw the ice queen in herself and sought a little preventive counseling, a dose of anger management. Bridgette admits she didn't inherit the effervescent, girly-girl capacity for sweetness, for chitchat, for saying "awww-uh" a lot, what she calls "the blah, blah, blah thing," like most of the women she knows. (Though she does envy it.) She doesn't like to talk forever about wedding plans, baby showers, wall colors. At some point, she will sit silently while Christine and Traci, her brothers-in-law's girlfriends, ably banter with Jeff's mother, Marie. "I just can't do it like they can," she says. Bridgette goes to her in-laws' house for every holiday except Christmas, and sometimes she clams up after the first couple of hours, when it seems like the talking will never stop.

Then there's a tender side. Bridgette can get weepy over Christmas ornaments and the collection of dozens of different snowmen that line the top of her kitchen cabinetry and are strategically placed all over her house. One of the first ornaments out of the box on this afternoon reminds Bridgette of her beloved grandmother, whom everyone called Meme. She died last year. Bridgette wipes a couple of tears from her eyes and keeps unpacking.

Bridgette is also one of those Christmas softies who collect little porcelain houses with tiny light bulbs inside them. She'll spend an entire afternoon unboxing her "snow villages," the collectible figurines and houses manufactured by a company called Department 56. It takes still another afternoon to connect her four dozen little houses to extension cords and plug the cords into surge protectors (that's the part Jeff likes). The little houses and sidewalk scenes go on the entryway console table and on stepladder-style bookshelves Bridgette has Jeff set up in the hallway and living room.

Snow village houses are still another American obsession at Christmastime, another way in which some believers pursue the comforts of something small and simple — for a price. Collectible snow villages first began in 1976 as a gift item for sale at Bachman's, a family-owned chain of floral and decoration stores based in Minneapolis. That grew into a massive collectibles company called Department 56, which is now headquartered in Eden Prairie, Minnesota. Department 56 describes its origins in the desire of a group of friends to perfectly recapture the sights and feelings of a snowy December evening they'd all shared one night at a country inn by the banks of the St. Croix River.

The company's first snow village set consisted of six miniature houses. Other models followed each year, becoming more intricately crafted and detailed, and soon included electric bulbs that gave the houses their warm inner glow. By the mid-1980s, Department 56 had unveiled a "Dickens" series of houses, shops, a church, and a town square, followed by a "New England" series, and then the "Alpine Village." These collections grew into hundreds and then thousands of pieces, complemented and populated by tiny Department 56 people, cars, trees, and municipal infrastructure such as bridges and streetlamps. Subsequent series collections glorified a quaint, midcentury New York of the mind ("Christmas in the City"), the biblical fantasy ("Little Town of Bethlehem"), and, most popularly, life with Santa at the North Pole. Early on, the company began retiring the original pieces, by reportedly breaking the molds from which they were manufactured. Such retirements became more frequent in the late 1980s, stoking that deep craving in certain consumers to hunt and appraise treasure. When a Department 56 mold is destroyed, the value of that piece may rise from an original retail price of $50 to $200 to several hundred dollars or more. A New York investment firm purchased the company in 1992. In 2005, Department 56 was acquired by a publicly traded housewares company called the Lenox Group. Snow villages contributed to $502.5 million in total sales for Lenox in 2006.

Hard-core Department 56 fans ("villagers") congregate on-line and at group meetings and swap meets; there are state and local chapters with bylaws and annual conventions, where they like to wear T-shirts and buy bumper stickers proclaiming themselves as a "Village Idiot." (According to Robert Frank's book *Richistan: A Journey Through the American Wealth Boom and the Lives of the New Rich,* the founder of Department 56 told the author that he used to rubber-stamp "Get a Life!" on gushing letters from village collectors. He called the most ardent villagers the "Get a Life crowd.")

Villagers cannot be typified as old ladies with an excessive Christmas jones; a majority of them were younger than fifty in 2001, according to the company's research. (There is considerable demographic overlap with model railroad nuts.) Once a happy villager owns more than ten villages — the average Department 56 fan reportedly owns at least two dozen pieces — there's an issue of where to store them for ten months a year. Department 56 was made for people with extra closets, garage space, and walk-in attics where they can store their villages in the original boxes, because true villagers never throw away the boxes. Soon enough a villager has a microcosmic concern with urban sprawl.

For weeks now, I've been learning about little houses (and the people who love them) by hanging out on Saturday afternoons in Holidaze & Gifts, a year-round Christmas store in a Plano strip mall. Holidaze is perpetually decked out in fat, nine-foot artificial trees covered in thousands of dollars' worth of Christopher Radko ornaments. It is a high-end heaven of angels and elves. There are four large tables of Department 56 Christmas villages that Holidaze keeps plugged in and on display. And it's not the villages I like to look at so much as the customers, who come into the store and lose themselves in snow village trances, deciding which new pieces to add to their collections.

One woman is getting ready to go to Oklahoma to set up her ailing mother's snow villages, just as soon as she sets up her own. A Plano obstetrician-gynecologist (and villager) is famous for the room-size snow village and model railroad display in his large home, a *Polar Express* replica that is featured in the December issue of *Plano Profile*.

One older couple, Lynn and Alan Wagner, former presidents of the national order of Department 56 village aficionados, invite me to their 8,500-square-foot manse in the Preston Trails neighborhood in Dallas, where they keep all their little houses on display year-round, in every room. The Wagners believe they possess one of the largest, most complete Department 56 collections of all time. It takes us an hour to look at them all. (Lynn dashes ahead of us to each room to switch on the lights in each display. "No peeking!" she admonishes. "We'll be good boys," Alan replies, and gives me a look, and I'm not sure if it's mutual giddiness, or if he's hoping I'll initiate his rescue by calling 911 the minute I get to my car.)

In her book *Merry Christmas! Celebrating America's Greatest Holiday,* Karal Ann Marling, a University of Minnesota art history professor, writes of the mysterious grip snow villages have on their owners: "Even if they never lived there, they remember the villages on Christmas cards, Jimmy Stewart running through the streets of Bedford Falls, places that exist only in the nostalgic realms of memory, imagination, and longing. Simpler times . . . The village lets the collector refashion this world according to his own wishes, remake her memories of the Christmases that should have been . . . [C]ollectors who have discussed their experiences publicly stress the magical ability to control reality or to create a fantasy that does it one better."

This sense of village making and controlled reality is a constant theme everywhere I go. Pieces of Frisco are being built all around me. Jeff is involved in a Christmas village creation on a large scale at Frisco Square, while Bridgette draws comfort from the small scale. One December afternoon in Holidaze, watching a twelve-year-old girl and her father stare with fascination at the sprawling Dickens village, I realize where I've seen that limitless gazing before. I recognize it in myself.

Not long after I first got to Frisco, I became similarly enraptured with a motion-activated flat-screen video kiosk in the City Hall lobby. It shows, on an endless loop, for as long as I care to watch, a computer-animated status report on the construction of nine new miles of the Dallas North Tollway. It's as mesmerizing to me as arranging pieces of Department 56 is to villagers.

It's the epic saga of Frisco, a PowerPoint presentation designed by superheroes. Imagine the aerial shot in the opening sequence of *Dallas*, only in reverse, zooming out from the city. Flying high across the Plano city limits and over the seventy-two square miles of Frisco, the viewer sees the new tollway in full splendor, set to a galloping, Aaron Copland–esque symphony meant to evoke pioneer optimism. A deep-voiced narrator proudly soars with us over perfectly rendered exit ramps and interchanges, pointing out all the progress — some of it already finished, about to be finished, and, the further you go, deliciously dreamed.

We swoop over Stonebriar Centre mall, the Ikea, the big-box power centers, the soccer stadium and sports arenas; over the computer-made images of sports cars and SUVs in the toll lanes below us. The narrator talks himself into a reverie about a mix of shopping, entertainment, civic community, luxury condos. (*When I first got here,* I supply the refrain of all Frisco citizens in my head. *When there were just cows.*) Still going, the narrator intones: "Local officials are obtaining right-of-way to extend the tollway northward again . . ."

I've stopped in the lobby and watched this video so many times that Mike Simpson, the talkative, outgoing mayor of Frisco, gives me a DVD copy to keep. He tells me of a recurring dream he had long ago, looking down on a future Frisco, the farmland turned into an assembled and orderly city, seen from the perspective of omniscience.

Bridgette collects only from the "Original Snow Village" Department 56 line, and her pieces include the snowman's house, Santa's Wonderland house, and a piece called "1224 Kissing Claus Lane"; she has the NASCAR café, the riverboat casino, a schoolhouse, and a Habitat for Humanity commemorative piece that shows a snow village house being built for charity. The tiny villagers in Bridgette's display have busy leisure lives, but they don't have nearly the shopping opportunities Bridgette herself has. Department 56 has never issued a Christmas world that accurately resembles our own — and who would buy it? Although there is a Department 56 Starbucks, there is no "box-store village" series in which to place that Starbucks next to the Chili's and the FedEx Kinko's, which could sit on zoned "pads" in front of a porcelain Super Target or twenty-four-hour Wal-Mart. There is no mall anchored by a Nordstrom, J. C. Penney, Dillard's, Sears, and Macy's. There is no Home Depot, Barnes & Noble, or Courtyard by Marriott. There is no tiny Tammie flying down a tiny Dallas North Tollway in her tiny Big Red filled with tiny tubs of tiny garlands. There's no tiny Caroll, Marissa, Michelle, and Ryan camped out in front of a Best Buy, waiting for it to open. In Department 56's world, the tiny Victorians and/or Bedford Fallsians are bundled-up, village-dwelling figurines who do not appear to need three-car garages in which to store all their Christmas decorations.

You can take that fantasy as far as you like: Department 56 does not have scenes of villagers arranging to drop off their children at the mixed-use condos of ex-husbands. No one is being grouched at by her mother-in-law or other relatives about who is deciding to spend the holiday where. Nor are the villagers drawing names to see who has to buy what for whom this year, in what exact color and what exact size. The adult villagers are not sitting in work cubicles e-mailing one another the explicit lists with Amazon.com links to what they want for Christmas. The villagers aren't buying gift cards out of desperation because they don't know what else to get the ungrateful nieces and nephews on their shopping lists. Instead, the villagers all tote perfectly wrapped parcels of presents, all of which are (this seems implicit) one of a kind.

A couple of nights before Merry Main Street, Jeff shows me his *pièce de résistance,* an extension-cord fettuccine special atop the white-pebble rooftops of Frisco Square, under a bright moon ringed by a shadowy corona, which illuminates trails of passenger jets crisscrossing the black sky. Later, Jeff checks connections to the lights wrapped around all the just-planted oaks that front City Hall. There is another glut of orange extension cords that lead to power sources underneath the building—spider holes into which Jeff happily crawls while I hold the flashlight.

"Is it on now?"

"No."

"Now?"

"No."

"Now?"

"Still no."

"Hunh."

To control the show, Jeff has installed a computer in the top-floor conference room space of Frisco Square's management corner office, wirelessly linked to all the circuit boards. Up here, he spends hours on spreadsheets detailing where each bulb connects to which board. (His brother Greg calls the office "Jeff Orgasm Central.") It would be easy to watch Jeff at his busiest and see a man singularly consumed, almost arrogantly confident in his own wiring. Alone, he can be witheringly funny about the incompetence of those around him, especially those whose job it is to install Christmas lights. With others, he is genuinely good-mannered (Eagle Scout; he calls every man "sir") and diplomatic in his disputes with the installation crews hired by the city and Frisco Square. When children come up to talk to "the man who makes all these lights," he humbly accepts their praise and answers their questions with basic facts of electrical engineering.

Jeff calls out numbers from his computer monitor and I read corresponding numbers back from a spreadsheet printout. We are searching for one miswired string that has gunked up a sequence. He heads back outside to check the Wi-Fi connection on the roof of the west building. I sit, waiting, and look around the walls of the conference room, which are lined with color-coded maps and drawings of Frisco Square's dreamers. Some of these drawings are already three years old. One envisions a Whole Foods grocery store, which seems unlikely now; another calls for a movie theater with a restaurant and pub (that deal ultimately falls through, too). There are plans for more chain hotels and bigger retail spaces and many more apartments and condos. There's also a sales-incentive poster on the wall that promises free trips and tickets to college bowl games for agents who sign tenant leases by the end of the fiscal year.

Jeff returns and taps some more on the keyboard.

"How's that look?" he asks.

I lean out of an open window and watch the real estate flash on and off, limned in white C-9 bulbs.

* * *

Christmas time! That man must be a misanthrope indeed in whose breast something like a jovial feeling is not roused — in whose mind some pleasant associations are not awakened . . . If your glass is filled with reeking punch instead of sparkling wine, put a good face on the matter, and empty it off-hand, and fill another, and troll off the old ditty you used to sing, and thank God it's no worse.

— CHARLES DICKENS, "Christmas Festivities"

Several thousand people do indeed show up for Frisco's official Merry Main Street at Frisco Square on Saturday night, December 2, in spite of a cold snap. They have occasional looks of genuine glee on their faces, alternating with puzzled inertia, as if unsure of the point or what to do next.

Why, it's for the children, of course: most adults here have a child (or grandchild) representing one of nineteen elementary school classes that are singing on one of three stages at each end of the plaza in front of City Hall. These simultaneous performances create an atonal duel of cross-chorales if you stand right in the middle. The kids from Curtsinger Elementary are on the main stage singing "The Twelve Groovy Days of Christmas," while "Jingle Bell Rock" is being belted out by Sem Elementary on another stage, and "Blitzen's Boogie" is being sung by Bledsoe Elementary on the other end.

There are lines to get into the gingerbread house exhibit. There are lines to buy pizza from Cub Scouts, get free hot cocoa from Genesis Metro Church, buy turkey legs from the First United Methodist men's club, and buy tamales from the St. Francis of Assisi Ladies Auxiliary. There's a long line for horse-drawn carriage rides around the parking lot. There's a line to meet Frisco firefighters and see the inside of their truck. There's a crowd around the North Pole elf who is making balloon animals. There's a line to get kids' faces painted. There's a line to meet Santa Claus at the Nicole Day Spa.

Parents everywhere are pushing double-wide strollers through the crowd, while others are in a constant state of documenting their children's evening with digital video recorders, desperate to make the footage become lasting Christmas memories. Parents ask their kids every few seconds if they're having a good time and get them to pose and tell it to the camera: *Smile, Tyler. Smile, Kayleigh. Smile, Isabelle. Smile, Jayden, Aidan, Jaycen, Keelan.*

The longest line by far — a two-hour wait — is for the Merry Main Street Kids' Holiday Store. Held in the largest of Frisco Square's vacant storefronts, the Kids' Holiday Store lets children wander through tables of merchandise (most of which is donated overstock from Target and a now-bankrupt sporting goods shop) where they buy gifts for their parents and siblings, at a cost of 50 cents to $5. There are basketballs, soccer balls, Legos, beer cozies, bottle openers with sports logos, boxes of golf balls, unread thriller novels, perfumes, shave balms, cheap jewelry, baseball caps, throw blankets, Polly Pocket dolls, nail polish, musical coasters, and tiny flower vases, among other stuff a mother or father does not need or want for Christmas, except on the heartbreakingly sweet level of the child's effort to gift them. (It's retail's earliest lesson: 'tis better to give than receive, and 'tis best to shop.)

The children are dropped off at one end of the line by their parents and then escorted through the store by one of Santa's college interns, out of sight from Mom or Dad. The kids "pay" at the "register" and then proceed to "gift wrapping," where another helper packs away their merchandise, to keep it a secret when they emerge from the other end, where the parents wait to retrieve them. "We got you a candle," announces one little girl to her mother, which leads her older sister to break out in tears.

"Why did you tell her?! I hate you!" she screams, throwing the bag with the candle to the floor.

The kids all stagger out of the Kids' Holiday Store with a look of exhaustion, having waited an hour or more to get in to pick over whatever's left of the merchandise and make their way through checkout and gift wrap. Here you have the entire story of retail Christmas in America, in make-believe microcosm: "Everything sucked," one boy says to his father. "You won't like what I got you."

"Did you find all sorts of good stuff?" one mother gushes to her son and daughter.

"I can't remember," the son replies.

Another mother sends her son back in, because he forgot to buy something for his brother, and the look on his face is close to utter despair.

By now, the mayor is on the main stage in front of City Hall, leading the crowd in a round of "We Wish You a Merry Christmas," before he counts down to the lighting of the tree. "Are we ready?" he asks. "Count down! Ten! Nine! Eight! . . ."

I turn and look across Frisco Square and up to the top-floor offices, where Jeff is waiting to start his light show on all the buildings. Bridgette and Greg and others are up there with him. When the crowd shouts "One!" the tree comes on—Jeff's cue to hit a button on his computer keyboard so all 150,000-plus bulbs of the square and City Hall will start their dance. Mariah Carey's voice comes out of the speakers. (When it all works, Jeff will tell me later, he is so overcome with relief he starts to cry.)

And just like that, it starts to . . . snow.

What I mean, of course, is that it "snows." Knowing that most days in December here will likely be sunny and temperate, the management at Frisco Square had snow machines installed at intervals along the square. It snows every twenty minutes, every night, bubbly bits of soap blown out with a loud whir, floating down. Kids start running back and forth, trying to catch ersatz snowflakes on their tongues, and every parent reaches for a camera. "Jasmine, don't *eat* it," a mother begs.

"What did you think?" Jeff asks Bridgette later.

"You want to hear my thoughts? Because they're not very nice," Bridgette says.

Jeff looks at me. "This ought to be good."

"It was cold as hell," she says. "I think it's retarded that the mayor led the people in a Christmas carol. The tree came on a second early . . ."

Jeff smiles.

"I married Madame Happiness," he tells me.

"I married Mister Christmas," Bridgette says.

10

Poverty Barn

"And how did little Tim behave?" asked Mrs. Cratchit.

"As good as gold," said Bob, "and better. Somehow he gets thoughtful sitting by himself so much, and thinks the strangest things you ever heard. He told me, coming home, that he hoped the people saw him in the church, because he was a cripple, and it might be pleasant to them to remember, upon Christmas Day, who made lame beggars walk and blind men see." Bob's voice was tremulous when he told them this, and trembled more when he said that Tiny Tim was growing strong and hearty...

— CHARLES DICKENS, *A Christmas Carol*

THE MAD SCRAMBLE is on to do better by others. Discreetly leaving a TMX Elmo in a Toys for Tots collection box is, in every way, the least one can do, and worse, it's far too anonymous. In these Christmas seasons of the *Extreme Makeover: Home Edition* era, when we're surrounded by the inspirational courage of Lance Armstrong and lectured by celebrities who arrive at eco-awareness concerts via private jets, no act of charity should ever be done in secret, the way the Gospel prescribes. Giving needs a good story line, too, a story that a corporate office can sponsor, thus exulting in the combined arts of marketing and beneficence. In this buildup to the week before Christmas, Frank Reed, the bottomless-hearted DJ at KLTY, seems to have increased the frequency of his on-air Christmas Wish granting, where the sobbing gratitude never stops: help for grandmas with autistic grandsons, help for working moms with broken-down Toyota transmissions, help for good Christian neighbors with nowhere to turn and empty gas tanks. Help is given with the added value of spreading positive PR for local advertisers. The KLTY van arrives at randomly selected Chick-fil-A restaurants for live promotional spots, to bestow blessings on the most deserving who are first to arrive.

We all do what we can. Tammie Parnell makes her mission trips to Mexican slums, and she has decorated the house of her pitiful, dying neighbor, and she is buying Christmas presents this year for one of Blake's former teachers at the Montessori school, a woman who is raising two daughters alone.

Jeff and Bridgette Trykoski set a cardboard box, colorfully wrapped in Christmas paper, out in front of their high-wattage house display each night, encouraging visitors to leave a canned or dry good to be given to the Frisco Family Services Center food bank. Frisco Family Services is Jeff and Bridgette's main charity, to which they write checks during the year. Dropping some of his collected food off at the center, Jeff is told that the food bank is out of ready-made entrées, and he runs out and buys several hundred dollars' worth of frozen pizzas.

Caroll Cavazos always buys a few presents to go to charity. Before Bank of America bought MBNA, her office would adopt a family for Christmas every year; Caroll used to be the first to volunteer to drive down to the south side of Dallas and deliver the food and presents to the recipients. "I loved doing that," she says. "You got to see it in their faces. This might be the nicest thing anyone had ever done for these kids, and at least they would have that memory. Everyone would help buy and wrap the stuff, but nobody in the office really wanted to do the delivery part, and I'd have to get four or five guys to go with me, to make sure everything was okay. I loved that more than just, you know, buying the presents for a family you never meet." Now her office has gone back to participating in an anonymous toy drive.

And what am I going to do for someone this year? Give what? How much? My dispassionate intent to merely observe has recently succumbed to the enchanting glimmer of the mega-Yule, and my Charles Dickens impulses draw me to the comfort and ease of something called the Angel Tree.

After another Sunday service at Celebration Covenant, I wait in the lobby for Caroll to go get Marissa from the kids' area. The sermon today made mention of how many of us in the congregation might be earning more than $100,000 a year, and everyone looked around at one another. (Could you mean me, Lord?) We were all supposed to envision ourselves as rich.

Across from the coffee bar, people are crowding around the Angel Tree.

It is a cheap-looking eight-foot artificial tree, onto which a few dozen white paper silhouettes of tiny angels have been tied with red ribbon. The angels are distributed by Frisco Family Services, each representing a needy client of the city's largest and only full-service charity. Each angel is labeled with a wish from a child or senior citizen living within the boundaries of the city's school district, the recipient identified only by age and gender ("Boy, 5," or "Girl, 11") and a specific request ("game for PlayStation 2" and "Heelys"; "makeup bag and lip gloss"). People are turning over each angel on the tree, looking for the one whose details they like best.

Caroll always lets Marissa pick out their angels.

"I always pick someone my age," Marissa says, "and I always pick a baby."

Our sense of Christmas is nothing without the narrative of heartbreaking need. Mary needed a place to give birth and nobody would give her one. This need for need exists so that our children can distinguish it from the concept of *want*. In the 1850s, when the American idea of Christmas was still nascent, Santa Claus was believed by white southern children to visit black slave children, underscoring the idea that *even they* got presents at Christmas. (Slave owners stood in for the concept of St. Nicholas, doling out small presents and days free from labor.) To fulfill the Christmas contract — a present for every child, even the poor — the myth of Santa has always called upon our ability to share with the less fortunate.

This leads to the permanently tender state of the American Yule, captured in the soft focus of Hallmark TV specials. It's difficult to exert too much criticism of an event that has such a conspicuously generous heart. We have much to tell one another about how deeply we care in December and, in turn, about how good that makes us feel about ourselves, which can be shared on talk shows, the news, and now on do-good blogs and YouTube videos (to say nothing of potential 501[c]3 tax deductions).

Christmas can always accommodate current fads in magical thinking. In the later part of this decade, the trend leaned (or regressed, for some cultural observers) toward telling people they can survive illness (especially cancer) and other obstacles with enough faith and positive thinking, or find financial success with nothing more than the resolve to believe it so. As Christmas of 2006 approaches, the new bestseller on the shelves is called *The Secret*, which essentially holds that one can outwit calamity by simply willing the negative away. To counter the selfishness of Christmas retail binges, we feed an enormous appetite for *Secret*-style, priority-adjusting anecdotes of the lost and found — houses burning down (and the occupants showered in blessings and old clothes), families living in cars (adopted, relocated to apartment complexes), people going into hospitals (an MRI scan coming back miraculously clean), long-lost relatives searching for one another (and reuniting in airports, by the Cinnabon). Some come straight from the newspaper, while others are only local rumors, snippets, Internet links. All of them are certainly wrenching:

The organ donor list is long, but the daughter holds on, and when her father speaks in her ear she squeezes his hand . . .

Witnesses said there were already-wrapped Christmas presents strewn across the intersection. Neither the driver nor the passenger survived, but the baby didn't have a scratch . . .

Toni's friends all wear matching butterfly pins to remember her struggle, one year after she passed on . . .

The tractor-trailer ran the red light, the visibility was low, the road was slick . . .

The search party is going out again into the Oregon mountains, while a Dallas wife and children wait and pray for their lost, adventuresome husband and father . . .

Few of these stories can end miraculously. (The mountaineers' bodies were found a few days later.) What they all have, however, is the frosty December veneer of a Christmas in need of need.

Everyday poverty is a much trickier story line, not just because it has a way of afflicting people in April, June, October, and anytime else far off from the charisma of Christmas. Everyday poverty is an especially jarring narrative in a high-income exurb because it requires a certain set of facts to line up for that tax-averse, conservative, evangelical voter opening his or her checkbook. The fictional Cratchits, in *A Christmas Carol*, are still the ideal poor family more than 150 years after we first met them. They are merely on a tight budget in a tough economy, and not unemployed, and do not go around asking for money or Angel Tree presents. This humility is what makes them so deserving. Bob Cratchit works as a clerk for Ebenezer Scrooge, in the Victorian equivalent of today's cubicle at a Countrywide bank branch in a strip mall.

Tiny Tim Cratchit's problem in 2006 is that he's rarely what you desire him to be, starting with the fact that he doesn't speak with a cockney accent, have cute coal smudges on his cheeks, or wear fingerless gloves. A few miles away, the city council in Farmer's Branch has joined a nationwide clamor for a crackdown on the vast number of Spanish-speaking Cratchits, passing an ordinance to prohibit landlords from renting to undocumented Tiny Tims and their families who've crossed the border to set about illegally mowing Texan yards, illegally caring for Texan toddlers, and illegally cleaning Texan houses.

It's best for everyone if the poor and sick meet society's Yuletide fantasies halfway. They must first believe fully in the spirit of Christmas. Their tales of woe, no matter how desperate, should have something to make the season bright: They will be cured. They will get a job. They will *get it together*. All they need right now is a Kohl's gift card, or that portable DVD player from Circuit City. The retail cure seems to me the easiest part. Like everything else, it involves lots of shopping, and in Frisco I find I'm getting better at that than I've ever been.

At a minimum, there are three ghosts of Christmases past, present, and future coming for the inner Scrooge that works alongside my snide comments and cool distance. I am long overdue for a Christmas that moves me to act in generous and optimistic ways. I need to pick some angels off that goddamn Angel Tree.

Thus motivated, I set out to buy gifts for four angel requests, which I plucked from various trees set up around town. The first two, which I took at Celebration Covenant Church's Angel Tree in the church foyer on November 26, have the code numbers C-541-1/D: 819 (the request is from "18, M, Gift certificate for shoes") and C-576-1/D:819 ("13, F, Gift certificate for J. C. Penney").

Angel Tree programs are in full force all across the land — operated by food banks and assistance centers like Frisco Family Services, or through churches and public schools. Angel Trees are more popular here than Toys for Tots, making the U.S. Marines' program, which dates back to 1947, seem like an antiquated gift-delivery system — especially when Toys for Tots cannot provide its donors the power to make specific wishes come true.

You claim the paper angels off the tree, fill out your name and phone number on the angel's lower half, and a volunteer tears it at the perforation and keeps that half for the center's records. The top half of the angel reminds you to turn in your gift (unwrapped, with the sales receipt if possible) either to the church or to Frisco Family Services' collection center, no later than December 12. Other than the gender, age, and wished-for item, there is no other information about who the child recipient is or how he or she got to be needy. This doesn't stop people from inventing elaborate stories about their angels. You can make them Hispanic, black, white. You can make them chronically poor, or you can imagine that they are just having a bad year. You can go totally Lifetime Original Movie with them and subject them to every problem and drama imaginable.

"It's a natural thing to do," Frisco Family Services' executive director, Jill Cumnock, tells me a few days later, as we sit in her office off Main Street in an old portable building that used to be a Catholic parish's offices. "People have this need, which I understand — they like to imagine who the child is, what they're really hoping for. I'm all for that — anything that raises awareness that there are people in our community, in this town, who need some very basic help, who can't afford the things we take for granted."

Jill's eight-year-old daughter, Abbey, got so excited shopping for Angel Tree gifts last year that she went to bed on Christmas Eve asking her mother all sorts of questions, wondering if their little angel had opened her presents yet. The Cumnocks had picked an anonymous girl about Abbey's age. Although it would have been possible for Jill to look up the code number on the angel and tell Abbey almost every detail about the girl and her family circumstances, she didn't want to betray clients' confidence, or spoil the fun. "Abbey would say, 'Do you think she likes it?'" They bought the little girl a set of Bratz bed sheets — depicting the big-eyed, midriff-exposing dolls, which satisfied Jill's social-worker need to get the child something useful and Abbey's need to get her something frivolously glam. "She had been thinking about this little girl, imagining what her life was like," Jill says. From a social worker's perspective, this is how you get people to start caring about the actual poor around them, and not just in December. You give them a story to hang it on. "We all need to know that what we're doing makes a difference, and that's what a story does for us," she says. "It makes us feel good."

Frisco Family Services clients in 2006 total 1,213 households, and the number is growing by about 35 percent each year, Jill says — a rate now faster than the overall population boom. The annual budget is $1.9 million. Some clients shop weekly at the center's food bank, where, based on household size, they get a set amount of dried and canned goods, day-old bread from a chain of bakery cafés, fresh produce, meats, dairy items, frozen entrées and vegetables, grocery store cakes and pies. Each month, the food bank typically gives away some thirteen tons of food, supplies, and toiletries. (In December, the amount given away at the food bank can increase by another four or five tons; not only does demand increase at Christmas, so does supply — for once.)

Other clients who don't need groceries instead receive financial help with their bills and rent, often because they've been laid off, fell ill, or aren't otherwise hip to *The Secret*. Some need a tank of gas to get to a job on the far side of Dallas or Fort Worth. Some receive a full Frisco Family Services spectrum, such as the 325 families (a total of 1,008 people) who fled Louisiana after Hurricane Katrina in 2005 and ended up within Frisco's school district in reduced-rent (or free) apartments and got food, job-hunting assistance, financial counseling, payments on overdue bills, school supplies, clothes, and Christmas toys.

Most of the Katrina people had moved on. Jill and her staff are back to dealing this year with an increasing number of what could be called the subdivision poor. These are clients who occasionally come to the food bank in late-model Ford Expeditions or Lexus sedans, which baffles volunteers who help load the grocery bags in the car. Jill is now seeing clients from neighborhoods she could never afford to live in. Name a subdivision — Starwood, the Plantation Resort, the Lakes of Preston Vineyards, the Trails — and she will nod and say, yes, there's at least one client who lives there. Some are struggling to make second- or third-mortgage payments on 3,500-square-foot houses, Jill says. The reason they drive nice cars is because they are caught in lease contracts or are too upside-down on the loan to sell the car.

It is common to hear people in Frisco talk about an increasing number of overextended credit crazies who are living in a fantasyland of pretend wealth. They loom large in suburban mythology, especially around Christmas, and are known everywhere as "the Joneses" (as in "keeping up with the . . ."). The cliché Joneses have been around forever. They exist mainly in anecdotal form, and they never seem to have specific names or addresses. It's always somebody *else* who lives like a Jones. I hear talk of them at church meetings, barbecues, and play groups, where they are known as "$50,000 millionaires." (The amount varied. They used to be $30,000 millionaires, but they seem to have gotten a raise.) They are house-poor white-collar couples and singles living with boggling (contemptible, to the person telling it) credit card debt. They have huge houses, but hardly any furniture inside except for (in the telling) the biggest, flattest, latest televisions. They drive luxury cars on leases. They throw big birthday parties for their kids. They shop constantly, their closets engorged with $900 handbags and $3,000 golf clubs. Mr. Jones is always screwing around with some other woman. Mrs. Jones is always inviting you to her Pampered Chef sales party. The Jones kids are always the kids who are peer-pressuring your kid on MySpace. The Joneses just got back from a shopping trip to New York or a cruise to Cozumel. They decorate their children's bedrooms entirely in Pottery Barn Kids, nothing but the best. It gets worse: The Joneses are rude to others in the Tom Thumb grocery store parking lot, where they park in handicap spaces without handicap stickers and never take their shopping cart back. In traffic, the Joneses won't let you merge from the tollway ramp—even as you signal! They occupy some psychic space next door, with the *Dateline NBC* sex predators.

The whole point to gossiping about the Joneses is to safeguard against turning into them, only to find we turn into them anyhow. "My sister is now one of our clients," Jill says. "She is the classic example of someone with Champagne tastes on a beer budget." Jill's father used to help her sister and brother-in-law out of financial troubles, until he died. Now it's the husband's father who's helping. "I have to stop myself from getting frustrated, because when he showers the kids with gifts for Christmas, I'm thinking, 'Why didn't he take that money and tell them to pay their mortgage bill?'" Jill says. "But why he did it is because people love bringing joy to kids. It's in their faces as they open up the gifts. And he paid for my nephew to have a really nice birthday party and I'm thinking a regular slumber party would have been just fine and that money could go to the bills. It's hard to think that way anymore."

Jill's job now is to acquaint Frisco's many Joneses with each other.

"It's about appearances," she says. "The big myth is that everybody is affluent and that there are no needs. For us, the struggle is 'It's not our neighborhood, it's not anybody we know.'"

By the second week of December, Stonebriar Centre is awash in opportunities for the consumer to access the warmth of holiday philanthropy — to shop while elevating the soul. Gap and several other retailers have joined Oprah and the rock star Bono with a wave of "[Product] Red" apparel and items — a $20 candle, a $350 jacket, $45 T-shirts — benefiting the Global Fund to Fight AIDS, Tuberculosis and Malaria. Zales sells $14.99 holiday bears to benefit the Make-a-Wish Foundation for dying children; J. C. Penney has a stuffed Rudolph to benefit its inner-city After School Fund. Bath and Body Works is selling candles to help Elton John fund AIDS research. Macy's has a line of baskets woven by Rwandan widows, to promote developing Third World economies. Starbucks is giving away 10,000 cards that say "Cheer Pass," which are supposed to encourage coffee addicts to perform an act of kindness for, well, anybody. Stonebriar Centre also has its own "Giving Tree" (owing to retail's secular aversion to calling it an Angel Tree), and if you pick an anonymous, needy recipient off the tree today, you are entitled to $5 off any $25 purchase at Build-A-Bear Workshop, a free Chick-fil-A sandwich with the purchase of a large drink and fries, 10 percent off at KB Toys, a free Great American Cookie, jewelry discounts at Claire's or Icing by Claire's, or a free sitting at Glamour Shots.

But I have my angels already.

In addition to my eighteen-year-old boy angel (C-541-1, "Gift certificate for shoes") and my thirteen-year-old girl angel (C-576-1, "Gift certificate for J. C. Penney"), I am now shopping for two more angels, which I'd picked off another Frisco Family Services Angel Tree at the Merry Main Street festival. For those, I'd looked specifically for angels in the AARP demographic, picking a sixty-nine-year-old woman (C-696-3) who had asked for a "gift certificate to restaurant"; and the angel who would become my favorite, an eighty-year-old woman (C-662-1) who had put "anything nice" as her wish.

Anything nice. Just as Jill Cumnock had described, I wander Stonebriar Centre four days before all Angel Tree presents must be turned in, shopping my own fanciful narrative musings about each angel.

The thirteen-year-old girl who wants the J. C. Penney gift card is, in my mind, one of three siblings in a Latino family that lives in one of the apartment complexes in Frisco near the Dallas North Tollway construction. Her father works on a landscaping crew and her mother cleans houses. They have one car, a 1992 Chevy Lumina. They get groceries from the food bank once every other month or so. She's a pretty good student. She gets online at the Frisco library and listens to reggaetón. When asked by her mother yet again to fill out her wish for the Angel Tree, she decides she can have more to choose from if she asks for a gift certificate to J. C. Penney.

"I'd like a $100 gift card," I tell a clerk at a cash register, and then I tell her what it's for and ask her if she thinks that's a lot, or just about right. "I think it's a lot," she says, pretending to even care. "I think that's a lot for someone you don't even know."

Good, I think, moving on to my next angel, the eighteen-year-old boy (a man, technically, if it please the court) who has asked for a gift certificate for shoes. But what kind of shoes? Should they come from some broadly bland but useful store — Payless Shoe Source? Wal-Mart? Or something fashionable from Nordstrom? Are we talking basketball courts, in which case Dick's Sporting Goods, Champs, the Finish Line? And by the way, what is he like? For some reason I'm hearing a lot of hip-hop dueling with Christian rock. He's in his senior year of high school. I can't tell if he has a part-time job, but for some reason I think he's busing tables. My mind makes him white. There's lots of hair gel, even though the popular boys are going shaggy now. His grades are awful. He hates reading. He's living with his mother, in a house that was built in 1990. He drives an old Hyundai.

This fantasy accompanies me to the other side of the mall, down near Santa's 'hood, to Journeys, which is a chain of shoe stores that rode a nice long wave of the Doc Martens grunge craze in the 1990s and has since tried to keep up via Skechers and Vans and the somehow still rebellious Chuck Taylor Converse sneakers. My angel could get some black shoes to wear to something like a job, or he could get hipster sneakers. He could even get Doc Martens. He could go goth or skater. I get a certificate in the amount of $150, enough for three pairs if one pair is on sale. I hope that works.

Focusing now on my sixty-nine-year-old lady, who has asked for a gift certificate to a restaurant, I'm having a hard time conjuring a suitable backstory here. She's not quite old enough to be living in a nursing home, so maybe she has come to the center's attention some other way. She's using the food bank, but why? Is she one of those grandmothers who winds up supporting her grandchildren? I give her a combined balance of $19,883 on a Bank of America MasterCard, a Chase Visa, and a Discover card, and I don't ask her why. A financial-planning counselor has her on a strict budget and actually took scissors and made her cut up her cards. At first she felt like she's too young to hang out at the senior citizens' center, but she does, and this is where she not only brushed up her bridge game, but also learned that she qualifies for food bank assistance. She has a few friends from church and would like to take them out to dinner. Exhausted by the banality of this particular fantasy, I decide to get her a $50 gift certificate to the Cheesecake Factory, since I'm here anyway, getting a bit drunk.

Finally, I have to do something about "anything nice," the gift desired by the eighty-year-old woman. I wander the length of the mall again and eventually drift into Brookstone, the gift store where customers are always sitting in vibrating massage chairs. It's all about stereo iPod ports and geosynchronous clocks. In the center of the store there's a large display of Brookstone's Nap line of throw blankets, thick socks, personal pillows, teddy bears. They claim to be the softest in the world, "created in pursuit of the perfect nap." I run my hand over all of it.

There's a lot of talk about the way electronics have changed our lives, but I think there's an argument to be made that the real consumer revolution has been in *soft*-wear. A product advertised as "beyond soft" or "softest" has this strange, synthetic allure. Your fingers try to forget about it but go on feeling it for many seconds after you've touched it. "One touch and you'll see" reads the ad copy on the back of the clear vinyl bag holding a tan ("latte"), fleecy, forty-by-sixty-inch Cuddle Dream Blanket, for $60. "The minute your skin comes in contact with a Brookstone Nap Dream Blanket, you begin to feel its therapeutic effects." I keep sticking my hand in the package and touching the latte blanket. Softness has become one of the American antidotes to fear, environmental crises, and the terror of the world outside our door. We became a self-cuddling people. Extreme softness probably is the nicest thing about Christmas stuff now, the lengths to which we've improved already-soft nighties, turned robes into wearable duvets, and given our plushy animal friends an eerily soft soul. Out of ideas, I accept that "anything nice" could mean anything soft. The only picture I get is a woman on a sofa dozing off to *CSI: Miami,* with the blanket pulled up around her legs. It feels the way lambs look in Nativity scenes — silky, off-white, furry little whorls of 80 percent polyester and 20 percent nylon. One latte Nap Cuddle Dream Blanket, sold.

And so, a few hundred dollars later, I have my angels covered. Now I wish to know if they are, in fact, anything like the Tiny Tims I imagined them to be.

11

The Neediest

J ILL CUMNOCK AGREES to let me observe behind the scenes — and help out, if needed — as Frisco Family Services Center's staff members and volunteers collect, inventory, and distribute the hundreds of gifts collected through the Angel Tree operation. Kimberly Girard, a twenty-eight-year-old social worker, is in her third year of running the program. Because the center has outgrown space at the food bank, the management at Frisco Square offered a vacant retail space that had briefly been a dry-cleaning shop.

Kimberly is standing behind the sales counter in the space on a sunny, seventy-degree Tuesday afternoon, December 12, wearing a red Santa hat and a tight T-shirt. She has an Excel spreadsheet open on her Dell laptop. She and another employee, Myrna Martinez, got extra-large sodas from the Subway next door and have the urban-hits radio station on, which is playing R&B holiday tunes from what already seems like many Christmases ago, by Destiny's Child, 'N Sync, and Boyz II Men. Behind a makeshift partition of hospital-blue sheets, they've set up tables and a work area where volunteers will retrieve and assemble gift packages for each client. The bare walls are covered with cardboard snowflakes and penguins.

Kimberly walks me through the operation. There are four

rows of temporary metal shelves, making narrow, twenty-five-foot-long aisles all the way to the back of the store. The shelves, which are seven feet tall, are already filling up with toys that have been labeled and arranged by code number. Clients with the lowest code numbers will begin picking up their gifts next Monday. The pickup appointments will continue on the quarter-hour, all day, in order, until the last client is scheduled to arrive late in the day on December 21. In theory, the shelves will empty out from right to left, then left to right, up one aisle and down another, until the last toy has been given away.

The gifts are double-checked in Kimberly's database before they're shelved. "People keep bringing presents that are wrapped, even though we tell them not to," she says. "We can't do that to parents. You don't want your child to be unwrapping something on Christmas morning and not even know what it's going to be. So the first thing we do is unwrap them." A fleet of new bicycles is parked along one wall. One child has asked for a giant red beanbag, which is next to the bikes. Along another wall are dozens of tubes of wrapping paper, which will be included in each family's bundle.

In the back storeroom, Kimberly sighs. "This is where all the extra stuff is going." It's piles and piles of new clothes and random toys, where the plaintive, electronic "bye-bye" of a buried Elmo can be heard from the bottom of the heap. Two volunteers are dividing the items by gender, which means half the room is turning girly pink and purple. The empty, rotating dry-cleaner racks are coming in handy here, for hanging up baby, toddler, and children's clothes by size and gender. It's going to take several days to sort through it all: "This is what people's generosity looks like," Myrna says. "These are all things that came in extra, on top of what we asked for."

Myrna, who is thirty, has long black hair and is dressed in tight jeans tucked into boots, and looks like she could be one

of Charlie's crime-fighting angels. She manages and guards a
file box that holds hundreds of gift cards, which she must sort
through and match up to client code numbers, as well as check
their validity on the retailers' websites. A lot of extra gift cards
have been donated this year, which will come in handy in another two days, when Kimberly will have a better idea of which
Angel Tree wishes didn't come back fulfilled. She'll then spend
one of the busiest shopping weekends of the year in the malls
and box stores, using gift cards to buy presents to make up the
difference. "One way or another, we're usually able to get everyone what they asked for, or close to it — either with stuff that we
had extra, or with a gift card," she says. According to her spreadsheet, there are 245 families receiving Angel Tree gifts this year,
with a total of 535 children. Another 39 recipients are senior
citizens. Each child submitted three items that he or she wishes
Santa will bring him or her, and each separate wish became an
angel, which was then hung on one of the many trees around
town.

I spend most of the afternoon with the other volunteers, log-
ging in each donated toy or other gift on the master list, then
arranging and shelving the items: "Girl, 6," has asked for and
will receive "anything from Club Libby Lu," the makeup store
for preteens; painting or coloring books; and a Tamagotchi electronic toy pet. "Girl, 9," has asked for an Imaginiff game, a Dora
the Explorer motion lamp, and any kind of board game. "Boy,
5," asks for any PlayStation 2 game, another PlayStation 2 game,
and "clothes." "Boy, 12," asks for a game compatible with Xbox
360 and any item featuring the Dallas Mavericks basketball
team's logo or players. A thirteen-year-old boy wants "T-shirts
with funny sayings." In addition to each child's three wished-for
presents, Kimberly says, it looks like Frisco Family Services will
be able to give every kid an additional three or four presents

from the surplus in the back room, for a total of six or seven gifts per child.

The volunteers are mostly stay-at-home moms in their forties. One woman brings her two daughters — a senior in high school and a freshman in college — so they can share the warmth of volunteering, but the daughters spend most of their time bickering with one another, eating pizza, and texting. There's a lot of talk about how nice all of this stuff is, how lucky all these kids are going to be. "This is better than my own kids do," another volunteer says.

Each day I spend at the toy drive, one volunteer always takes it upon herself to boss the rest of us: *You're filling out the sheet wrong, the checkmarks go here . . . Don't start a pile there, I've already started sorting those . . . Don't give them that wrapping paper, I was saving that for a family with a lot of girls because the snowmen have pink scarves — give them this one, instead.*

As we work, I keep inhaling the intoxicating air of new clothing and plastic toys sealed off in clear plastic clamshell packaging. I notice, with some deflation to my sense of originality, that a lot of the thirty-nine senior citizen angels will be getting super-soft blankets. I had dropped off my presents a few days earlier, and so I start reading each angel tagged on the shelf, looking for either my sixty-nine-year-old woman who's getting the Cheesecake Factory certificate, or the eighty-year-old who's getting the Brookstone Nap Dream Blanket.

I stop searching because a group from a church arrives in three large SUVs to unload several dozen Angel Tree presents, including two more bikes. Many of the gifts are wrapped in pretty, coordinated paper with Scotch invisible tape and tied with ribbons and bows. After the church people leave, behind the sheet partition, our afternoon becomes some odd parody of Christmas morning, as two of us sit on the floor and unwrap the presents without ceremony, then mark them down on an inventory list and wedge them onto the shelves. If they don't match a code number or they come back lacking proper angel ID, we gently toss them atop the pile in the back.

A little past 4 P.M., long after the volunteers have left to pick up their kids from school, Kimberly and Myrna realize they're late for an appointment. Kimberly locks up and gets in her car, Myrna gets in her car, and I get in my car, and we speed together four miles away to the Hall Office Park, a collection of glass and granite boxes interspersed with sidewalks, fake lakes, and new trees. We park at an office plaza on Internet Boulevard.

The event today is both an in-house Christmas party and a public-relations opportunity, where Frisco Family Services' toy drive organizers (that would be Kimberly and Myrna) will meet and chat with some of the employees of Skywire Software and iWave Integrator, two tech firms that have adopted the Angel Tree program as their seasonal charity. Inside a large meeting room, about twenty-five employees, many of them obligingly wearing red Santa hats, are sitting at tables assembling the contents of several dozen Christmas stockings that have each been packed with small toys (Matchbox cars, faux-Barbies), candy, and kid-friendly toiletries such as Dora toothbrushes and Anakin Skywalker shampoo.

Jason McDonald, iWave's marketing manager, is in charge. He's an affable, chubby-cheeked cubicle character. He asks someone to go upstairs to the office and see if they can round up more employees to come down, lend a hand, and pose with Kimberly and Myrna for a group photo. There's also a buffet — employees brought in Sam's Club sugar cookies and made a peach salsa sour-cream dip for chips. There are sodas and DoubleStuff Oreos. Someone made divinity.

Jason explains that employees were asked to contribute, if

they wished, $20 to the cost of filling each stocking. They fell nine stockings short of their goal of a hundred, so the CEO "did the others out of his own pocket," Jason enthuses, spelling the CEO's name for me, which I make a show of writing down. The company asked some employees to bring their children and spouses in, so Helen Pitts, a woman in a festive red sweater vest who is in charge of corporate PR, can take more group photos of the assembling of the stockings, and then get photos of the stockings being handed over to the Frisco Family Services employees. These photos will be a good way to let people know, as Helen says, "when [the company] is doing something that helps the community."

"Now these are all labeled, right?" Myrna asks of the stockings, looking the finished ones over. "Like, boy or girl, and age?"

Oh, yes, we are told; each stocking has a Post-It note, see? ("We'll sort them out when we get them back to the storeroom," Kimberly whispers to Myrna, worried about the sticky notes falling off.) She and Myrna cheerfully go out to the lobby, where Helen says the light is better for pictures.

"They were nice," Kimberly says, after it's over.

"Weren't we supposed to load up the stockings and take them with us?" I ask, thinking that the point of racing over here in three cars was to pick up a hundred stuffed stockings.

No, Kimberly says. They just wanted her and Myrna there for the pictures.

The next afternoon, the stockings from iWave and Skywire are indeed delivered to the Angel Tree distribution center, where volunteers have hung them on the old dry-cleaning racks, along with other stockings assembled by employees of other nebulously named high-tech companies in other office parks. But when it comes time to distribute them, in the rushing around and cramped quarters behind the blue sheets, volunteers (myself included) will have a difficult time finding the right stockings for the right ages or genders.

More often than not, we will wind up dumping a stocking out for a quick scrutiny of its contents, to see if it is appropriate or useful for that particular child. If not, we grab another stocking, dump it out, and try again, until finally we cobble together and restuff the right kind of stocking. Kimberly will keep saying, "Don't forget about the stockings," but we forget about the stockings every time, and I will chase bewildered families out to the parking lot to explain, sometimes in my remedial Spanish, that they are getting *los "stockings" tambien, uno momento.* In families with more than one child, we'll have no fast way of labeling the stockings for each sibling. Whatever Jason McDonald and his cubicle mates had imagined about bringing special stockings into the lives of specially impoverished children, what really happened was this: after the last family has come to pick up its Hefty bags of toys and wrapping paper tubes on December 21, we are left with several dozen stockings — many of them from Skywire and iWave — and Jill Cumnock, her daughter Abbey, Abbey's best friend, and I will sit on the floor and dump them out, sorting toys and other gift items (which then went to Frisco Family Services' thrift store) from candy and toiletries (which went to the food bank).

Ever the social worker, Jill will have only passing regret about this. "It would have been nice if we had organized this part better," she said. "The good news is, it all goes to people who need it. It's not wasted."

Sometimes the neediest are the ones doing the giving, those of us who need charity to fit some fantasy we're having about our own spiritual well-being. Sometimes the neediest are the ones who have plenty, like the corporations seeking publicity for quasi-mandatory employee charity. The givers can tell themselves whatever story works best. The reality is a pile of Dora toothbrushes, destined to brush someone's teeth anyhow.

I find it curiously comforting to watch as these Tiny Tim notions give way to the genuine gratefulness in the faces of Frisco's poor. As volunteers finish sorting and cataloging the thousands of toys and other presents that are ready to be picked up by client families, I double-check entries on the master list of code numbers. Since Kimberly has data about the clients right there in her laptop (and because she knows almost all of the clients personally), I'm hoping she can share with me, generally, some details about the very angels I'd shopped for a few days earlier. This way, I can see if what I'd imagined about them is anywhere near the truth. I may even ask Jill if I can arrange to meet them, and perhaps spend part of Christmas Day with them, to follow my act of giving all the way to the act of receiving.

But as I search the shelves and cross-check the master list, I notice that the children's code numbers do not extend as high as the numbers I'd drawn off the church Angel Tree — C-541 for the boy who wanted shoes, C-576 for the girl who wanted a J. C. Penney gift card. And my numbers for the elderly women are much higher than what is shelved and waiting, although I notice the Brookstone Nap Dream Blanket is here with a neat pile of remaining, unspecified gifts on the lowest shelf.

"Oh, that's because they aren't really people," Kimberly matter-of-factly says when I ask about the code numbers.

Come again?

Kimberly then explains an essential tactic to staging a successful Angel Tree drive: "We add extra things that clients and their kids might have wished for, in case we don't get everything that they actually did ask for." This is because many angel wishes get lost, or aren't fulfilled, or are filled with the wrong item. Frisco Family Services always creates a couple of hundred more nonexistent angels to add to its trees around town, giving them ages and genders and specifically *non*-specific wishes ("anything nice") to reduce a generosity gap. Many of these extra wishes are for gift cards ("J. C. Penney," "gift certificate for shoes"), which will help with the shopping spree the center does for its actual angels, or will come in handy for future clients' needs.

I fell for exactly what the Angel Tree does best: I opened my heart and visited the mall and let my imagination and my Visa card make up the story of whom I was helping, which turned out to be nobody real. "No, it's a good thing," Kimberly says. "You'll see . . . everything goes to someone who will really appreciate it."

This explains the mountain of toys in the back room: they are all surplus, even though many of the families who have donated them are probably operating, as I was, under the idea of a specific child of a specific age who had put each toy on a wish list. It took three volunteers most of two days to sort through that pile alone. The Bratz are all eyeing me, seeming to scoff at my realization. I turned out to be the neediest one of all.

Sympathetic to my journalistic need to make paper angels into real children, Jill has one of her caseworkers at Frisco Family Services arrange real interviews for me with a few of their real clients, so I can learn more about what led them to be on the receiving end of the Angel Tree.

I spend some time with Elizabeth Brzeski, a forty-four-year-old single mother of four (two girls and two boys, ages five to eleven), who lives in a three-bedroom apartment in Frisco, barely scraping by on a $30,000-a-year job at an accounting firm. Elizabeth drives a burgundy, repair-prone minivan. She and her ex-husband once owned a software start-up. They moved to Texas from Silicon Valley sight unseen during the tech boom, mainly because the husband had read about Collin

County on a list of the fastest-growing U.S. technology corridors. Loosed from the high price of California real estate, the couple bought a 5,000-square-foot house behind the gates of Starwood. They'd lost it all in 2002 — the business, the house, the marriage. There was a time when Elizabeth bought whatever she wanted at Stonebriar Centre; now she avoids the mall (truly) at all costs. She can't afford to shop. A sweater set that she saw in Nordstrom the other day triggers a self-pity she cannot bear.

In an interview she gives to a local TV station, for a minute-long story with "falling from the upper class" as the holiday-woe angle, the reporter adds a new, entirely inaccurate twist to Elizabeth's story: her house had burned down. (It had not. The bank had foreclosed on the mortgage.) "I didn't understand that," Elizabeth says, after the story airs. "Like, 'She lost everything'?" We puzzle over it. Is the story not miserable enough on its face? The lowered income, the ex-husband who is chronically late with child support checks, the cramped apartment? Elizabeth wonders if "house fire," rather than simply being inaccurate, is the reporter's attempt at bleak euphemism.

Another Frisco Family Services client — a newly unemployed fifty-four-year-old wife, mother, and fibromyalgia sufferer named Denise Matise — literally becomes the poster child of the invisible, middle-class food bank recipient. Frisco Family Services is part of a twenty-two-agency, citywide, $1.8-million *Dallas Morning News* charity drive, which had asked for real, live clients to feature in an advertising blitz (on billboards, in quarter- and half-page newspaper ads, radio spots, and TV commercials). Denise's face was now all over town, under the words *I am a survivor* in a bold font. At first, it troubled her to see her picture everywhere. "But I had to give something back," Denise says the first time I meet her. "I am sometimes just stunned how much we've been blessed by God and the people around us."

Denise, a faithful evangelical Christian, is always in a state of enumerating and giving thanks (in her New-Yorker-turned-Texan accent) for small mercies. Her working-class poverty comes with middle-class contradictions, such as broadband Internet access and a DVD collection. Denise lives in a two-bedroom apartment she shares with her husband, Hollis, her fifteen-year-old daughter, Danielle (a strikingly pretty honors student who goes online as "jesusdork" and stages youth awareness rallies for global aid to Darfur), and a terrier named Angel. Hollis worked in the high-tech sector until he was laid off five years ago. That first Christmas he lost his job, the family received assistance from a group of tech-savvy Samaritans calling themselves Geek Santa. After a fruitless job search in the field, Hollis now works the night shift as a trucking dispatcher, thirty-five miles away in Mesquite. He drives the family's PT Cruiser there and back and spends his days in an apnea-prone doze on the couch. (Because of Denise's fibromyalgia symptoms — which include debilitating fatigue with muscle and joint pain — the couple has not shared a bed in many years; Denise's bedroom is heavily populated with collectible Victorian baby dolls, which she occasionally sells on eBay for extra income.)

Denise has milky soft skin and spiky hair dyed a punky shade of plum (she's a frequent customer at the discount Toni & Guy hairstyling academy in Carrollton). She is days away from going on Social Security disability pay, having thoroughly exhausted every sick-leave option from her part-time job as a legal clerk. A single day of work — parking at a light-rail station in Plano, riding to downtown Dallas, and then walking six blocks — usually confines Denise to bed for the next two days. She says she has to take a lot of prescription painkillers. Fibromyalgia still carries a stigma, since not all doctors agree on what it is or even if it exists.

Denise's favorite escape is her bathtub, surrounded by candles and a variety of lotions, soaps, and dolls. Denise and Hollis are not the inert needy, but they are often exhausted trying to pay their basic bills. Hollis plays guitar in their church's praise band, and the family has managed to act as hosts for foreign exchange students. Danielle is currently sharing her room with Jun, a teenage girl from Japan. One can only guess, when talking briefly to Jun, what she thinks of her year in America . . . in the richest suburban county in Texas . . . surrounded by Mc-Mansions . . . to wind up living in a walkup apartment where financial instability is a daily issue, where the price of gas influences whether this week's groceries will come from Wal-Mart or the food bank.

Yet the Matise household works hard to present a picture of mirth each time I visit. Denise's doll-themed Victorian Christmas tree could rival a Tammie Parnell production. On the morning of December 19, with the air outside like a damp, gray towel draped over the world, I pick up Denise at her apartment for her pre-holiday trip to Frisco Family Services, to get her groceries and the girls' Angel Tree presents. Denise is wearing a long faux-leather coat, leopard-print scarf, and dangly earrings and walks with her cane. On the way, she looks out at the fog on the undeveloped pastures and says it's the most beautiful morning, so misty and ethereal, and it makes her think of Christmases back east. She is indefatigably upbeat today, even as she talks about how nervous she is that her unemployment checks won't last until Social Security kicks in.

At the Angel Tree gift center, she waits in a chair while Kimberly Girard's volunteers assemble her family's presents. When they come out from behind the blue sheet with two full Hefty bags of gifts for Danielle and Jun, Denise is moved to tears. "It's too much," she says. "Really, it's too much." (In my car, after we leave, she means it — *too much*. "The girls didn't ask for all this stuff." They had asked for gift cards to places like Kohl's and

Wal-Mart and Hot Topic — which are in the bag — but the largess of Angel Tree has also provided other items, such as two portable CD players, a hair-straightening iron, and cutesy makeup bags. Denise is going to give the surplus to her church. She's regifting angels.)

Next, at the food bank, Denise gives me a useful tutorial in making the most of what happens to be available. Amid the usual oversupply of donors' unwanted cans of green beans, pumpkin-pie filling, and hominy, Denise scores such rarities as bakery cakes, extra-virgin olive oil, and Hello Kitty bubble bath.

Driving back to her apartment, Denise and I are talking when, on the radio, KLTY plays a Christmas Wish. "Turn it up, I love these," she says. "Do you listen? Oh, that one the other day, about the mother with the two boys who was walking to her job because the car was broken? Danielle and I were driving somewhere. I had to pull over I was crying so much."

After I unload everything, Denise and Hollis insist I return in a few days, for the Italian sausage and pasta feast she is making on Christmas Eve. The Matise family will eat and watch DVDs all night. *The Polar Express, Christmas with the Kranks* — and we always, always watch *Christmas Vacation*," Denise says, trying to entice me. "Please come!"

In spite of straitened circumstances, both Elizabeth Brzeski and the Matise family members remain active participants in the seasonal economy. Elizabeth was overjoyed to find boxed sets of lotions at a Tuesday Morning store marked down to $9 each, which she was able to open at home and divide into gifts for her coworkers, pressured to do so by the approach of an office gift exchange. Denise happily discovers an extra $50 in one of her final paychecks and spends it at Wal-Mart, where she gets Danielle an MP3 player that is on sale. Hollis has secretly bought Denise a new television for her bedroom, which she does

not yet know about. Denise ordered size 36 khakis from Lands'
End for Hollis but is miffed to receive a rain check instead, telling her the pants are out of stock. Still more merchandise is going on Denise's credit cards, an extravagance in every way, but
she has decided she just doesn't want to worry anymore.

12

The Total Moment

They do not believe except what they see. They think that nothing can be which is not comprehensible by their little minds. All minds, Virginia, whether they be men's or children's, are little . . .

— "Is There a Santa Claus?" *New York Sun*, September 21, 1897

TAMMIE CALLS AND I GO, back into the land of plenty: Santa Claus is making his annual appearance at Stonebriar Country Club's Sunday brunch, and seven-year-old Blake Parnell — with his blond hair combed and parted, wearing a handsome red sweater vest — has come ready to sit on Santa's lap and present him with polite requests on a trifolded sheet of looseleaf.

"Dear Santa," Blake begins, and moves right to his wants:

A dirt bike.

A knife.

A PSP with movies ("Pirits of the Carbean" he wrote) and an "odgrafed piccher" (autographed picture). "Your the best, love Blake," he concludes, and draws a picture of himself and Santa holding hands. "PS — Sorry I mest up."

"This Santa is *phenomenal*," Tammie tells me in a hush, while we watch Blake and the genuinely bearded one smile for their professional photograph. "We've had him here before. I hear he's really hard to get." In Tammie's world, it's important to be careful even about choosing the Santa your children encounter. Someday Emily and Blake won't believe in Santa, a prospect that propels Tammie into a wistful state as she considers the irretrievable loss of innocence. She worries that *this may be the year* for Emily, her ten-year-old. Parents here trade tips, in online forums, about when to tell your child *the truth,* and how to delay the trauma another year or two. They call in to radio talk shows and ask for advice. They envy other parents who have children who still believe. They resent other people's children for spilling the truth at Indian Princess meetings or at sleepovers. A couple of days ago, while we were in her kitchen, Tammie told me a great story she had just heard, about a little girl who'd written to the editor of a newspaper asking him whether there really is a Santa.

"You mean 'Yes, Virginia'?" I said, and Tammie nodded.

"Have you heard it?"

Tammie, alert to every permutation and style of anything having to do with Christmas, has somehow missed "Is There a Santa Claus?" otherwise known as "Yes, Virginia" — the classic 1897 *New York Sun* editorial penned by Francis Pharcellus Church in response to eight-year-old Virginia O'Hanlon's letter to the paper asking if there really was a magical St. Nicholas who brought toys while children slept on Christmas Eve.

Tammie has a vague recollection of her own "No, Virginia" moment. It was around the time she was Emily's age, maybe ten or eleven. She doesn't remember specifically how it happened.

But Emily *knows,* yes? Emily knows that I know she knows, and I know what you might be thinking, and no, Virginia, I didn't tell her a thing. One afternoon, we're working together at the Parnell kitchen table to make a tiny, pagoda-esque cricket cage tree ornament for her book report on *A Cricket in Times Square*. (Tad's still at work, Tammie's on a decorating job, and the ornament is due in two days.) Emily tells me Santa exists, then pauses and looks at me ever so slyly with her mother's eyes, which leads me to privately conclude that she's doing the absolute kindness of keeping the fantasy alive for her parents and little brother.

Tammie is protective of Christmas in general. As she slogs through her final decorating jobs in the second week of December, I begin to see her work as an epic conservation effort — the guarding of preciousness. She tries to keep her children away from multiple, substandard Santas in malls to avoid Blake's increasingly skeptical questions. (Emily has stopped asking.) Tammie also keeps her kids away from the mall for other reasons: "Too easy to become a statistic," she says. For the presents that the Parnell children give to their teachers, Tammie now likes to set up a mini store on the family's dining room table, featuring various tree ornaments or other unused, regiftable items she's accumulated, or jewelry and other trinkets from various trade shows and bazaars. She lets Blake and Emily circle the table and "shop" for their teachers, tutors, sitters, coaches, and others.

"If Emily asks me if Santa Claus is real, what do you think I should tell her?" Tammie asks me, one afternoon when we're alone, in another house that is getting the garland-on-the-staircase, feathers-on-the-tree, full-on Tammie treatment.

I am ever a reporter and, it's important to underline here, *not a parent*. Virginia O'Hanlon was eight when she asked her famous question, and that seems to me like a fine age to get just a bit more real. I had my own "No, Virginia" moment when I was seven, after Christmas Eve Mass with my family. As I was escorted, half-asleep, from the car to my bed, I overheard one of my big sisters — Ann, always the loudest — telling my mother that I was dead asleep and it was okay to start setting up Santa's unwrapped gifts to me under the tree. I shed not one tear.

As a treasured piece of American journalism history, the full text of "Yes, Virginia" fails upon further scrutiny, if only because its ultimate message is that there is something inherently wrong with skepticism. If a child has concluded, all on her own, that it's impossible for a man in a flying sleigh to make it all the way around the world in one night, delivering elf-made replicas of all the stuff you see in Target and Best Buy, then that's a child I would be happy to steer toward a voting booth when she's eighteen. That's an American in search of facts. If, however, she goes on pretending to believe well into her teens (I encountered more than one such teenager in Frisco), because it makes her parents (and God) feel sweet and happy, then I become worried. That becomes an American willing to spend $100,000 on her "special day" wedding, or who will believe without hard evidence that other countries harbor weapons of mass destruction when they don't. The angst over Santa's existence comes not from the children, I think, so much as the grownups. The adults literally tear up when I ask them to talk about how, and when, their child will learn there is no Santa. Once you know there is no Santa, then there's no stopping the awful truth about everything else.

I do have good advice here for Tammie, which I can't claim as my own, since I read it on my smart friend Nancy Nall's blog. Nancy has a daughter a bit younger than Emily, who asked for and got the straight and simple answer, but with a twist. "Tell Emily the truth," I say, "but what you have to do is tell her that now that she *knows* about Santa, it's her turn to *become* Santa. Isn't that the point? That there's a part of Santa in each of us, and once we know, we get to 'be' Santa, and do something for some child, our own child someday, who still believes?"

This almost works; I can see Tammie considering it. She would gain a helper Santa and lose a little girl.

"I don't think we're ready for that with Emily," she finally says with a sigh, and we set about installing garlands on yet another staircase.

When Tammie first moved to Stonebriar Country Club Estates eight years ago, she hung out with other moms who had young children. They'd get together for Bunco or nights out for drinks or afternoons at the club pool. Somewhere along the way, the women started calling themselves the Hotties. (Actually, it's "Gold Hotties," which is an acronym — GOLD HOTTIES — that stands for Girls Out Laughing and Drinking; Husbands Often Trying To Imitate Every Supermom.) The Hotties used to see each other all the time, but husbands got transferred and one family goes relo, then another. Bigger houses opened up in bigger neighborhoods, with more square footage and larger yards. The Hotties try to keep in touch, even the ones who've moved out of Stonebriar.

Tammie and her friend Kelli, another Hottie, go over to decorate Lorraine's house for Christmas. Lorraine's also a Hottie. "We're some of the last of the original Hotties still in the neighborhood," Kelli observes. Lorraine's house is a block from Tammie's, around the corner and down a cul-de-sac. Her breast cancer has recurred and cannot be fought this time. In addition to round after round of chemo, Lorraine has done all that a twenty-first-century cancer "diva" must now do: she keeps an online journal; she did the walks and runs for the cure; she raises awareness, keeps friends close, and keeps God closer. "This is the real, horrible deal," Tammie says. Lorraine's husband works as a vice president at the headquarters of a global soda manufacturer; they have a daughter and a son. All season long, Tammie and Kelli insisted to Lorraine that they would come put up her Christmas stuff, in whatever way she wanted it done. Tammie and Kelli tell me that when they got there, on a Thursday night in November, Lorraine was propped up on the couch. She had on a styled wig, her makeup was done, and she wore a cute outfit. "Heels," says Kelli, amazed.

Lorraine sat and watched as Tammie and Kelli unpacked the ornaments and decorated the tree. She advised on the mantel decorations, the greenery, the Nativity scene, the dining room table centerpiece, and the stockings on the fireplace. Tammie knew this was goodbye. She wouldn't think of taking Lorraine's offer to pay for her services. "I said to [Kelli], 'We are staying until this house looks absolutely phenomenal.'" And when they had, Lorraine said she liked it, even as she sleepily drifted in and out of awareness.

Tammie has promised herself she will drop by to see Lorraine at least once a week. But as December goes by, she gets busier and drives past the turn down the cul-de-sac to Lorraine's house every day.

At one of the last houses on her client list, Tammie and I are in the kitchen, trying to decide what to do about the cookie jars and little Santa and Mrs. Claus salt and pepper shakers, and then we're in the family room, rearranging the mostly bookless bookshelves with an array of Christmas bric-a-brac that replaces the rest-of-the-year bric-a-brac.

"Hey," Tammie suddenly asks. "What's *going on* with this thing in Iraq?"

Between "Please Come Home for Christmas" by Don Henley and the Clay Aiken version of "Mary, Did You Know?" KLTY has just played another Christmas Wish, and this one has to do with a wife whose husband is in Iraq. I spaced out during the wish segment, gazing outside at the gray afternoon and the eerily silent weekday world of nice neighborhoods. But Tammie listens: the wife gets a laptop with a Skype connection, so her kids can see Daddy on camera while they talk to him on Christmas Day.

Tammie has apparently sensed a shift out there. President George W. Bush's approval rating has never been lower. In the November midterm elections, the Republicans have just lost the Senate and the House majorities, which seemed almost unthinkable two years ago. The Iraq war is — by any measure except that of the Bush administration — a disaster. There's also the feeling that other parts of our world are coming unraveled. The mortgage foreclosure rate in North Texas, for example, is higher in 2006 than it's been in the last decade, and rising. Why is that? No one will yet say. Tammie hears news in bits and pieces. Until now, the extent of our conversations about politics had meant only that Tammie made sure that I knew she was a friendly Republican, big on personal values. I had made sure that Tammie knew I was part of that East Coast media elite she'd heard about, with all the horrors that may entail for her. The result is apolitical; we get along like a house afire. "You work for a newspaper. What's Iraq all about? Help me understand it," she says. "I just never have time to read up on it, to read the news. I'm not up on the whole deal. So much has happened, you know? What's going on?"

"Um, the whole war?" I ask, not sure where she wants me to start. Should I dial back to the first Gulf war? Pick it up at 9/11? Explore whether Tammie, like 57 percent of Americans, thinks Saddam Hussein is linked to al-Qaeda? Talk about Saddam Hussein's pending execution? (Mary, tell me what exactly you *do* know, and I'll try to work from there.) "Well, it's essentially a civil war now," I begin, with the briefest possible explanation of the Sunni and Shiite factions. I describe how American troops are like a police force that cannot control crime.

"Well, how do we get out of that?" Tammie asks, now fiddling again with the kitchen table centerpiece. She's gone with a big, bright red, feathered cone. "Fabulous. Look at that. It was born for that."

I then start to get into some of the scenarios for a gradual withdrawal and bring up the Iraq Study Group's recommendations just delivered to the president, which, according to National Public Radio the other day, the president apparently has chosen to ignore.

But Tammie's gone. A minute or two on Iraq is as far as I get. "Where do you think these guys go?" she says, holding up some ugly reindeer figurines.

We return to the bookshelves, and now she's putting some little angels on the mantel. "What do you think of that, too many?" she asks.

I tell Tammie I'm lately all about taking things away. Two angels instead of five. One elf instead of three. A gradual withdrawal of troops.

"Minimalism!" Tammie gushes. "I know you're all about the minimalism."

A while later, Tammie tells me Tad is mad at her. She came home exhausted from yesterday's house, her hands scratched up even more. There are circles under her eyes. She occasionally nibbles rice cakes while she works and not much else, and she's lost twelve pounds since November. Blake is behind in his reading assignments. Tad says this has got to be the last year for this Christmas decorating business. The conversation always leads down the same path: What is Tammie trying to do here, what is she out to prove, why does a small-time decorating business have to take over their lives? What does Tammie want?

Tammie was thinking about it all night. What did she really want? (To be a great mother; to be a phenomenal business-woman; to return to corporate America.)

Which leads us to another discussion: What does Tammie want *for Christmas*?

Jewelry's nice. She also wants the complete DVD boxed set of *Alias*, the television series starring Jennifer Garner as a secret agent who dons a variety of disguises and knows martial arts. It was canceled last spring, and it was Tammie's only show. (She normally hates watching TV and has a special contempt — more like pity — for almost all celebrities and the deplorable political views she is certain they hold.)

Now Tammie is thinking about what she really wants. "Hmmm."

Mrs. Marshall, Emily's fourth-grade teacher, had the students solicit handwritten Christmas memories from family members and then assemble them in a book. Both Tammie and her older sister, Dee Dee, independently wrote of the same memory from Christmas 1972, in Eustis, Florida. That would have been around the same time Tammie figured out the truth about Santa. It was also the year she and her sister desperately wished for horses. Once the presents were opened, all Tammie had was a small plastic horse. Then her parents opened the drapes to the backyard, and there were two horses grazing by the fence, a palomino named Tonk for Dee Dee and, for Tammie, a pretty pinto named Penny.

Tammie's been looking for what she calls "the total moment" ever since. The horse was a total moment, and she had another one years later, in 1988, when she took a boyfriend to meet up with her family on a Christmas trip to Munich. She remembers the falling snow the night they got there, Christmas Eve, and the little church in a picture-perfect German village, and the complete stillness.

The next day, in one of the biggest houses we've been in yet, Tammie announces: "I've had a total epiphany! I have to tell you about my epiphany."

I hold up my notebook, indicating I am ready to receive.

"It's not about the stuff," she says.

I look around us. Here we are in a house with zebra rugs and a grand piano in the white marble living room. We are about to risk breaking our necks getting the garland and poinsettias on the foyer chandelier. The woman of the house has removed all photographic traces of the husband who recently left her. ("Divorce," Tammie whispers, the minute I arrive, as the nanny removes the young child from his Froot Loops breakfast and Nick Jr. cartoon reverie in the media room.)

It's not about the stuff?

"That's where I was going with this," she says. "I was thinking — all these people are paying me thousands of dollars to decorate their houses, when they could be doing so much more with it, to help people. Am I promoting that whole materialistic thing? How am I doing good? I mean I'm blessing people, I know . . ." But? "I just suddenly realized that, oh, my gosh, no — I don't want to buy Blake that dirt bike. Those things are dangerous."

If there's no dirt bike, Tammie may never hear the end of it, because Paxton Pearson has one. The Pearson family is a constant presence in the lives of the Parnells — lake houses, nights at the country club, carpool duties. Tammie and Tad had been talking to their friends Rick and Claire Pearson about taking all the children to Steamboat Springs, Colorado, with both families skiing and lodging together. But then Tammie hesitated, thinking it's too expensive — that as much as they enjoy being with the Pearsons, what she really wants is to be with Tad and the kids and no one else. "The total moment," she keeps saying. A dirt bike is not her idea of a total moment, and while she's at it, she'd also like to banish electronics — the PSP, the Dance Dance Revolution, the iPods.

So, at the last minute, Tammie has bought four plane tickets to Denver, spending several thousand dollars of her decorating business earnings to take Tad and the kids to a ski lodge at Vail, Colorado, for five days. They'll leave the same afternoon school lets out for Christmas break, on December 15, and get back to Frisco on December 20, in time for Tammie to barely finish the rest of her gift shopping.

"How will Blake take it when he finds out he isn't getting that dirt bike?" I ask, thinking of all the times Blake has assured me that Santa is bringing him a dirt bike.

"That's the thing. I've got to make it seem like something even better is happening. Wait'll you see what I have planned," Tammie says. "Have you heard of Cookie the Elf?"

The only time Tammie and the other Hotties get together anymore is for a Christmas party and mystery-gift swap. This year it's at Tammie's house, and she insists I come, as a sort of honorary Hottie. "Don't forget — there's a gift exchange. It's something fun we do, not extravagant. Bring something fun! Like $25 max, got it?"

Wednesday night, twenty-four hours before the Hotties party, Tammie has finished her last big client of the season, in a 6,000-square-footer in Starwood. She has of course pronounced it phenomenal: "The nicest, nicest people — my best customers."

On Thursday, Tammie is exhausted. She feels sick. Earlier in the day, she threw up and then took the only nap she's admitted to in the time I've known her, springing up an hour later to hie thee to Costco for the Christmas wassail: she buys ready-to-bake hors d'oeuvres, cookies, fruit, wine, and a chocolate fountain. "It says you just plug it in . . . you heat the chocolate . . . help me figure this out," she says, opening the box.

By the time the Hotties begin arriving, Tammie has given her house the good Christmas glow. The tree in the living room is picture-perfect, as is the garland on the mantel. She's outdone herself on a dining room centerpiece of angels and the three Wise Men, along with an array of snack trays. Christmas creatures gaze over the family room; she's decorated the kitchen buffet with lighted glass boxes made by a woman she met at 'Neath the Wreath. ("And she has the greatest story about how she fought cancer, and—" Tammie begins, but gets distracted and doesn't finish.) In the downstairs guest bathroom, I pull back the shower curtain and discover the only evidence of life getting ahead of Tammie Parnell — the bathtub is piled high with Easter baskets, plastic jack-o-lanterns, empty vases, old frames, unfolded towels. (*Dear Santa, I can explain.*)

Tammie's neighbor Marty Sue is the first Hottie to arrive, and she opens a bottle of wine while we wait for the others. Blake and Emily are hovering in the kitchen. Tammie told Blake he'd have to stay upstairs and wait for Tad to get home, but she's given Emily special permission to stay for the gift exchange portion of the party. Blake has learned that his best friend, Paxton Pearson, is leaving for a special ski trip with his family to Steamboat Springs, as soon as school lets out tomorrow — the same trip he knew his family had considered going on. He's sad and jealous. This is all according to Tammie's master plan — let Blake think the Pearsons have it better. Let his envy brew. Without knowing, Marty Sue is helping to stir Blake's bad mood when she starts telling Blake that she and her husband and son are leaving for Utah to go skiing for Christmas, and how much fresh powder was in today's ski report, and how her son will be snowboarding in twenty-four hours — "as soon as school lets out." Blake sulks off upstairs for a PlayStation binge.

Kelli, Renée, Jolene, Helen, Polly: one by one the Hotties arrive, eleven women in all, wearing their cute new tops, hair done, still trying after all these years to dote on and impress one another. *You look great. No, you look great. How are you so skinny? I miss you. I can't believe it's already Christmas.* The chocolate begins gurgling and circulating in Tammie's fountain on the buffet. "Tammie," Kelli says, "are you sure you don't want to put something down on the carpet under this?"

The Hotties are keen to any potential disaster now because a few years back, Tammie set her house on fire during the Christmas party.

"Not on *fire* fire," Tammie says. (*Dear Santa, I can explain.*) Tammie had made a beautiful arrangement out of real greenery and magnolias and lit candles in the living room fireplace and on the mantel, just as she had seen in one of her magazines, just like, you know, the nineteenth century. "Never again," she says.

Kelli remembers hearing a crackling sound from the living room, where Blake, then a toddler, was playing under the tree. Marty Sue remembers the burning smell and instantly knowing it was bad. All the women turned toward the living room to see flames shooting out of the greenery toward the ceiling. Marty Sue moved the sofa and started trying to smother the blaze. Tammie lunged for the gilt-edged formal photography portrait of her and Tad and the kids that hangs over the fireplace. ("That's a $3,000 picture," she wants you to know.) Another Hottie started trying to rescue the wrapped presents from under the tree. Kelli called 911. Helen kept Blake and Emily calm outside. The house filled with smoke. The Frisco fire department arrived in a few minutes and neighbors came running out of their houses. The room was doused with fire-extinguishing foam. Embers and smoke filled the house, but the damage was minimal. "Thousands of dollars," Tammie says.

Telling the story now gets everyone laughing, except Tammie.

"It wasn't funny at all," Tammie says.

Not long after the flames were out, Tad — the always calm, gentle Tad — drove up to the house and yelled at Tammie in front of all the Hotties. When the fire trucks left, Tammie walked to Marty Sue's, where some of the other women had gone to continue drinking. "Let's just put that one behind us," Tammie says tonight.

On the coffee table in the family room, the Hotties' pretty, pretty presents for the mystery exchange put the lie to Tammie's assurances that the gifts would be no big deal: they are crisply wrapped, or contained in glittery gift bags bursting with coordinated tissues, feathers, and bangles. "Presentation is the biggest part of it," Tammie says. The gifts I've brought are last-minute and larky: I bought yet another box of Snow Powder from Eitan, the Israeli army veteran at the mall, and, from Urban Outfitters, a cardboard Christmas tree that promises to expand and grow when you add water, sort of like a *ch-ch*-Chia pet. Both of these I wrapped hastily in brown paper with a cowboy Christmas motif and no bows. ("So *man*," Tammie says.)

Settled into Tammie's plump couches and chairs in the family room, the women are sufficiently buzzed to start the gift swap. You draw numbers. Whoever gets number 1 gets to pick the first gift (any of the presents, other than the one she brought) and unwrap it in front of the group. Number 2 can either steal number 1's just-opened gift or open one herself. Number 3 can now choose to steal either number 1 or 2, but no gift can be snatched more than three times.

There's a *Playboy Video Centerfold* tape that one of the Hotties found in her husband's closet years ago that keeps reappearing every year to predictable squeals. ("Emily, don't look," Tammie says.) Similarly, there's a tacky white ceramic wedding cake topper that always gets shrieks of dismay and is then regifted the next year. Renée unwraps a garish-looking glass bowl with a candle dish that floats in the center of it. "I brought that, I'm sorry," Polly tells her.

"What is it?" Helen wants to know.

"It's really just a big margarita glass, maybe," Marty Sue quips.

They fight over a pink pair of pajamas with matching slippers from Target. There's a decorative wicker cone with ornaments and a candle holder. Tammie has given a little cardboard gingerbread candy shop with a small sachet pillow inside reading "Dear Santa, I want one of everything." My Snow Powder and just-add-water tree turns out to be a popular item; several of the Hotties try to swap for it to take home to their kids. Marty Sue has gifted a nice bottle of Muscat dessert wine (I swap for that). Another Hottie brought a bottle of Pinot Noir in a cylindrical wine box that looks like a *Nutcracker* soldier.

By now the Hotties are tipsy, screaming and laughing at stories about Christmases past, about people they know. The economy is cruising. The war is distant, barely there at all. The house isn't on fire. The wine is tasty. "Let's toast," Tammie says.

"I think we all know who's not here this year," Marty Sue says, taking a serious tone. "And let's just take a minute to think about her."

"Have you seen her?" Jolene asks.

"We decorated her house," Kelli says, pointing to herself and Tammie.

"It was hard," Tammie says.

"But you should have seen her, though," Kelli says. "She sat on that couch and watched us do everything. It's so important to her. She got up, got dressed, perfectly, of course, and made sure to check out everything we were doing." The Hotties want more details. "Oh, those poor kids," Helen says. "How are they doing?"

You can tell, Tammie says. You can tell it's almost time.

You can tell how the daughter moves through the house and hangs back, Kelli says, how much fear there is, what death in a big house feels like, with friends and relatives coming and going.

"To Lorraine," Marty Sue says, raising her glass.

After the last of the Hotties leave the house around midnight, Operation Cookie the Elf is set into motion. Tammie and Tad spend an hour secretly packing some of the kids' things — snow-suits, ski jackets — for the big surprise tomorrow. Tammie has hired "Miss Cookie the Elf" to come over after school and reveal the news to Blake and Emily that Santa Claus has decided to send the family to Vail.

Tammie picks up Blake and Emily from school at one o'clock; Tad is there when they get home. ("Dad got off work early to-day," she explains.) Cookie's arrival is imminent. Tammie is hoping Cookie will be fun and energetic, but not scary. She's given Cookie a script to follow. She wants a big reveal. A total moment.

Miss Cookie the Elf arrives in a late-model red Volkswagen Beetle. She is a woman in her early fifties, dressed in green-and-red lederhosen, striped stockings, and green Doc Marten boots that curlicue at the toe. She wears a red hat, pointed ears, a red ball on her nose, and a lot of rouge. Her rate is $150 an hour. The problem with Cookie — I can sense right away — is that she's almost as shy as Blake and Emily are. Her voice is barely a whisper.

They let Cookie in.

Toby the dog gives her a security sniff.

Cookie invites the kids to sit on the couch with her; Tad grabs his camera, and Tammie sits across from them. Cookie explains that her main job is baking and decorating cookies at the North Pole. Then, in December, when it gets busy, Santa sends her on "special errands." She explains that the Volkswagen is a rental, and that somehow she gets here from the North Pole on one of Santa's "shuttle flights" but always falls asleep on the way. Cookie gets out her report and a letter she says is from Santa:

"'Dear Emily and Blake' — that's you," Cookie says. "'We've been keeping an eye on your from our control center' . . . And so we have every boy's and girl's name in the system. In fact, we've got your entire family in our file. Did you know that? You have to be very, very careful what you do through the year," Cookie says. "Are you good through the year?"

"I think so," Blake says.

"Mom and Dad?" Cookie asks.

"We work on it," Tammie says.

"Are you doing good in school?"

"We're working on it," Tammie answers for Blake. "We love to play sports and sometimes it's hard to sit down and focus."

Cookie follows Tammie's requested script:

"'We know that your mom has a very special job this season, and she has spent all her time at others people's houses making sure that they will have a good Christmas just like you guys, doing the trees and the garland and the mantelpieces and the Nativity scenes, making sure they're really beautiful. But did you know that sometimes when she's doing that, she talks to the little Santas and the elves?' Can you imagine someone talking to all the little porcelain Santas and the elves in the house? Do you ever catch her talking to them? How about you, Blake?"

"No."

"What she doesn't know, though, is that these little porcelain elves and Santas and snowmen, they're listening to her! One day we heard her worrying out loud about how little time she has to spend with you guys and how little time she would have left to make your Christmas special . . . You know what she realized? It's not about all of this stuff. Is it? It's not about all the presents — it's about being with one another that's really important, and it's about being with ones that you love. If people didn't have one another, it wouldn't be a very nice Christmas, would it?"

The kids shake their heads warily. Where is this going?

"So, she wished out loud for something very special," Cookie says. "Time."

"I need lots of time," Tammie says.

"So guess what? Santa heard her," Cookie says, and here comes the whammy: "So he decided that the best present this year was NOT a laptop for Emily, NOT a dirt bike for Blake, but that the best present right now for this family is to go on a fun trip! To make some great memories!"

Silence.

The dog cocks his head at Cookie.

Blake is processing the words *NOT a dirt bike.*

"Blake, do you like the mountains and the snow?" Cookie asks. "What about you, Emily? At Christmastime? It's really beautiful, isn't it? And you like to ski and snowboard? Well, back at the North Pole, we looked at your mom's wish and ran it through our computers, and the elves' executive committee has decided that the Parnells are going on a ski trip to Vail, Colorado!"

Blake and Emily stare at her, emotionless.

Cookie seems on the verge of a mild crying jag instead of hyper merriment. "Are you excited? Have you been there before? You're going for five whole days. Are your bags packed?"

Tammie tries to rally the room, laughing. "Yes!"

"Your flight to Denver leaves at 4:55, so you've got to hurry!" Cookie says. "You have worried looks on your faces. It's for five days and you'll be back in time for Christmas, you will! So you're going to spend Christmas here. All right?"

Tammie gets it now. It's too much for her kids to process, and it doesn't address a key concern: "Santa won't forget us here," she says.

"You guys better hurry, you have to be at the airport in an hour," Cookie says.

Finally Blake speaks up: "I have a question."

Cookie: "Okay."

"How did you get here?"

"Well, like I said, I come down on a shuttle, and they drop me off at the airport, and I rented a car. Because they won't let us use the reindeer and the sleigh until Christmas Eve, you know, because it makes all the control towers at the airports crazy if they see all these extra objects in the air. Now, let's get you off on your trip —"

"Can I tell you what else I want for Christmas?" Blake asks.

"Why, yes, you can."

Blake feels it's absolutely essential that Cookie pass word to Santa that he needs the new Xbox 360. He's already lost a dirt bike in this.

"And what else do you want, Emily?" Cookie asks.

"Uh, Dance Dance Revolution Extreme 2," Emily says.

"You guys want some really high-tech stuff," Cookie says. Blake keeps pressing her. Don't forget *Pirates of the Caribbean* for the PSP, he says. The *second Pirates* movie, not the first.

"I don't know what the elves are making over there in the high-tech division, but I'll see what I can do," Cookie says. "I saw they were making a dancing doll. Do you want a dancing doll, Blake?"

"No, ma'am."

"What about you, Dad, what do you want for Christmas?"

Tad, who's been photographing his children's stunned reaction to this bizarre woman, is caught off-guard. "Oh, Cookie, I don't know," he says. "Look, I just want everybody to be happy and together. Doesn't that sound like fun?"

"Now," Cookie says, "I'm going to give you each a candy cane and I want to leave you with the story of the candy cane. A candy cane is a letter *J*, see? And it represents the precious name of Jesus, who came to Earth as our savior. It could also represent the staff of the Good Shepherd. The candy is stained with three small stripes," which Cookie says represent the "scrounging" of Jesus. (Scrounging?) "The stripes represent the blood," Cookie goes on, and —

"Hey," Tad interrupts, "we've gotta catch a plane to Denver, right?"

Cookie gets up. "I'm going to give you guys hugs and let you get packed! You be good. Shake a leg, but don't break a leg!"

But Blake now has lots of skeptical questions. "How long does it take you to get back to the North Pole?"

"Twenty-five hours," Cookie says, headed for the foyer.

"Have you ever given someone a snowmobile?" Blake asks. "Are those your real ears?"

"All right, we better pack," Tad says. "Goodbye, Cookie! You tell Santa hello for us!"

The odd elf walks out to the street, waves, then gets in her VW and drives away.

"Get upstairs, guys!" Tammie commands. "Go, go, go. We have to be at the airport in one hour. Hurry, hurry, hurry."

"We're really going?" Emily asks.

They're really going. It's a flurry of rolling luggage, and changing the shredded nest in Mandy the hamster's cage, and making sure the house sitter will have what she needs. Blake insists on calling Paxton Pearson to tell him that Santa's elf just came by and has sent the family to Vail. "We'll call him later," Tammie says. "Hurry."

So much for that total moment.

"She was sort of too quiet, wasn't she?" Tammie says about Cookie the Elf. She hefts her rolling suitcase — filled with her pashmina shawls, Ugg boots, and "Born to Shop" socks — off her bed and onto the floor. "I thought Cookie would be more energetic. My kids are already kind of quiet around strangers. I think they were a little overwhelmed."

The dog has seen the luggage and started sulking, pacing about, his jingle-bell collar jangling despondently. Tammie leaves me a list of two things to do. One, she forgot to deposit the mortgage check — will I mind driving through their bank and dropping it off for her? The other thing is the FedEx man — he'll be here any moment with the harp, Tammie whispers. That's Emily's other big present this year: a harp that once belonged to an aunt. Tammie has had it restored by a craftsman in Minnesota. Tammie can't wait for Emily to learn the harp. Will I mind staying until FedEx comes?

A few minutes later, the Parnells are backing down the driveway, on their way to D/FW and toward Tammie's total moment. They will ski Saturday, Sunday, Monday, and Tuesday and be back in plenty of time for her to Christmas-shop for Blake and Emily. She'll still have time to gather her wits and see Lorraine before she dies. Tammie waves goodbye to me from the front seat as they drive off.

The dog and I look at each other.

13

Hallmark

The message is that the consumer is clearly alive and well
and supporting economic growth.

— MICHAEL P. NIEMIRA, chief economist, International
Council of Shopping Centers, in *USA Today*

WHETHER THE GLOBAL Christmas economy is looking up or looking down on the nightly financial news,
it will always have Bridgette Trykoski on the case, and untold
millions of Americans like her, eternally willing to make a trip
to a mall. Jeff doesn't quite get what she enjoys so much about it
(what is there to buy in the world, except for Christmas lights,
extension cords, and office supplies?), but one thing he knows:
"I like to watch her in the mall," he says. "She flutters. There is
absolutely no logic to it. I would like to be a little fly that can
read thoughts and just fly behind her."

What Jeff sees as random seems, to my eye, much more sci-
entific, even precise. It's late on a Sunday afternoon in mid-
December and they are in Stonebriar Centre, with just a few
things left to do Christmas-wise. The lights on the house are
under control, and the lights at Frisco Square are running

smoothly (though Jeff is always on call, ready to drive there at a moment's notice to deal with bulb outages or a Wi-Fi network issue). The inside of the house is decorated with Bridgette's precious snow villages and snowmen. The Trykoski siblings and Jeff's parents have all drawn their Secret Santa names and e-mailed their gift requests to one another.

The mall is placid this evening. We go from one end (Macy's) to the other end (Sears) in a matter of minutes. Bridgette is a finely tuned shopping machine. "Your mom needs a bear," she says, zipping through the Hallmark, and what about her mom, what about those kitchen thingies?

"I thought we were done with your mom," Jeff says.

"Honestly, you never pay attention to anything," Bridgette tells him. She moves immediately to another display. "Like a butterfly," Jeff says. "See?" He gazes at her with true amazement, and she turns to him and scowls.

"Jeff! Shut *up*," she says. "Focus."

"Happiness, happiness," he says, giggling.

This is one of the only times I've seen Jeff and Bridgette by themselves. (And even now, there's a dude following them around with a notebook.) So much time has been spent putting up lights, and people are always coming and going — Jeff's brother, Bridgette's friends, Bridgette's brother and his wife and their toddler. There was a big house party last night, to celebrate Bridgette's thirty-first birthday, with lots of the Trykoskis' Aggie friends. And, of course, there are always cars full of people idling in a line down the street, waiting to park and stare at their house.

But tonight it's just the two of them and the mall, and for once they are not in a hurry to make anything happen. As Jeff teases Bridgette, and as she grouches at him, I finally get them as a couple. ("It took me forever to get them as a couple," Greg's girlfriend, Christine Meeuwsen, confirms for me, when I ask people, including Jeff and Bridgette themselves, what makes that marriage work.) As he's following his pretty butterfly around from store to store, I glimpse it in Jeff's twinkling eyes: he's wild about her. She can step all over him and he loves it — and she loves him for loving it, perhaps? Later, in Staples, while he swivels happily in an office chair he wants to buy, spinning like a child, Bridgette notices something that displeases her: "Jeff, you blew out the crotch of another pair of jeans?" she demands to know. Mentally she adjusts her Christmas list: new jeans, always the same pair, always stonewashed blue.

"What were we, Lutheran first? Then Methodist?" This is Jeff's brother Greg talking, when I ask him what flavor of religion he and his brothers received growing up. "Our mom was Catholic." Mostly they didn't go to church and don't now. Greg's girlfriend, Christine, has tried out a couple of local churches, some of the really huge ones. Bridgette doesn't like big churches. "I hate the holy-roller stuff," she says. Bridgette grew up with the Assemblies of God, in the church with red carpeting where she and Jeff were married six years ago. "I just like it simple," she says. "Some traditional hymns, and then the minister gets up and reads something and talks about it. That's all." In truth they belong to the church of multicolored light strings, which preaches the spreading of good cheer. They believe in God, though — there's a manger scene in their living room, and they don't like seeing *Xmas* or hearing *holiday* where the word *Christmas* ought to be.

Jeff's parents, Jack and Marie, drive three hours from their home north of Houston to Frisco on Saturday, December 16, for what has become the great Trykoski family Christmas compromise, around which the tension never fully subsides, even in summertime: since Jeff and Bridgette are on their fourth year of staying put for the holiday, Jack and Marie make a visit a week or so before Christmas so the Trykoski brothers and their significant others can exchange presents as a family.

The Trykoskis start this evening off with good seats at a dazzling afternoon matinee concert of the Trans-Siberian Orchestra at the American Airlines Center in downtown Dallas, with a light show any careful epileptic would do well to avoid. This is followed by a drive back to Frisco and dinner at the Italian restaurant in Frisco Square, where Jeff gives his parents a full tour of his and Greg's 150,000-bulb musical monster.

After that, the family goes to open presents at Jeff and Bridgette's house, which has been flashing its own light show since 6 P.M. (the computer starts it automatically) for traffic that now extends three or four blocks around the corner. Inside, once the coats are off and beers have been opened, Jack sits in the overstuffed chair by the tree. Doug and his fiancée, Traci, are on the couch. Greg is on the recliner and Christine takes a seat on the floor in front of him. Jeff and Bridgette sit on the floor, too, where it's easier to pass out presents. Marie is up and down, antsy to go outside and smoke a cigarette.

The Secret Santa method leaves little to chance, precisely how most Americans like to give and receive their Christmas presents. Some of the earliest Christmas newspaper ads of the nineteenth century implored shoppers to "get them what they really want," and ever since, gift giving has carried with it a special neurosis of causing (or trying to avert) disappointment. This perhaps best explains the 200 percent increase in sales of gift cards in the last seven years; Americans seem to accept that they no longer know what to give one another.

Present opening can't start until Greg goes into the guest room and "bachelor-wraps" the gifts he's giving. Bridgette serves Jell-O shots in lemon, peach, and blackberry. The cat finds a spot by the gas fireplace. The XM radio network on the big-screen TV is playing carols by the Jackson 5, the Beach Boys, and Sarah McLachlan.

"Are we ready yet?" Marie says, once Greg emerges.

"Pick a name at random," Jeff says.

"Traci," Marie says.

Bridgette has drawn Traci, who unwraps a book of quotations about literacy (she's an elementary school librarian and teacher), and then a Texas A&M platter and A&M dishtowel. "Uh, did you go to A&M?" Greg jokes.

"It was on my list," Traci says, then coos one of those long, girly *thank you*s in Bridgette and Jeff's direction.

"Then Traci picks who gets to go next?" Marie asks.

"That will cause a paradox," Jeff announces.

"Whatever, Jeff," Bridgette says. "Whoever's name Traci has is next."

On it goes: Traci has Jeff's name and gives him a new backpack — "for when you go back to school," she says, referencing Jeff's upcoming MBA night classes. "Do you like it? Does it work?"

Jeff has Greg, who gets a set of torque wrenches from Sears and the real showstopper: a portable spotlight with a beam diameter the size of a medium pizza ("17.5-million candlepower" it says on the box). "You can go get wild hogs with that," Christine tells Greg, knowing her boyfriend fancies himself a kind of backwoods warrior. The three brothers rush out to the backyard to shoot a beam of light at the sky. Bridgette takes a Jell-O shot break in the kitchen, then shouts, "Y'all get back in here. We aren't done."

It's Greg to Bridgette now: he got her candles with snowmen on them, a Banana Republic gift card, and a Department 56 house — the Dairy Land Creamery from the Original Snow Village collection. Doug has Christine (two fleece jackets and a boxed DVD set of the first season of *Grey's Anatomy*). Christine has Doug (a case of Miller Lite, a box of cheese snacks, and the first season of *The Sopranos*).

The presents that Jeff and Bridgette bought and wrapped for Jeff's parents will go back home with Jack and Marie, to be opened on Christmas morning, as is Marie's preference. Last year, Marie was so mad at Jeff and Bridgette for not coming home that she never sent them the gifts she had bought them, which instead sat out in the bunkhouse behind Jack and Marie's house for months. ("Did we ever get them?" Bridgette wonders.)

"Well, I want to watch you open this," Bridgette pleads to her mother-in-law, picking up a large box wrapped in red and gold paper and handing it to Marie, who consents to open just this one: a set of hand-painted wineglasses. "I've seen these," Marie says, brightening. "I've coveted these. She knows I love all these pretty things."

Bridgette seems relieved.

"Thank you, Jeff. Thank you, Bridgette," Marie says.

"This is a wedding present," Greg says to Traci and Doug. "Or an engagement present. It's for Traci. Well, I guess it could be for Doug, too, but not really."

"I couldn't hold off anymore," Christine says, handing Traci the small, silver-papered box.

"Whaaaat?" Traci wonders, unwrapping it. Suddenly she's sucking in her breath as she recognizes the sparkly thing inside the small box. "It's my headpieeeeeeeeeeeeeeeeeeeeeeeeeeece!!!" she screams. She then jumps off the couch to hug Christine and then jumps around the living room waving — what? What is it? Hold still! — a small, glittery headpiece that pins to the veil she'll wear for her wedding in June. She'd picked it out at the bridal shop but hadn't bought it, had fretted about not getting it, worried it was gone. She can't stop screaming for joy, hyperventilating almost, as Marie and Christine loudly supplement the detailed narrative of how hard it was to keep it secret, going back and buying it, trying to figure out when to give it to her, and so on. Traci's mother died last year, and Marie has actively stepped in to help, as a surrogate wedding planner.

"Okay, she's happy," Jeff says. Traci is over at the dining room mirror, with the headpiece atop her blond head, loving it, her cheeks bright red with glee.

There's something deeply satisfying about watching people open Christmas presents, and taking notes on it, at a remove. It may seem terribly ordinary. Yet, as anyone who has ever sat in a group of family and loved ones and opened presents surely knows, there is an allure in watching humans unwrap and accept the banality of material objects with the occasional hysterically happy response. Absent the grandchild Marie so clearly desires and rarely mentions, the room focuses wholeheartedly on Traci's joy over the bridal headpiece. Moments like this make a family a family.

This would pretty much have been my report on pre-Christmas with the Trykoskis, if it weren't for the war in Iraq. It's another kind of moment, a darker one, which also makes a family a family.

Christine and Greg have been dating for four years now, and no bridal headpiece seems to be in her immediate future. Greg's not asking, and she's not demanding. It's been a sad year. Christine's younger brother — Bill Meeuwsen, a handsome, twenty-four-year-old, newly wedded army sergeant — was killed in a gun battle two days before Thanksgiving 2005 on the outskirts of Baghdad. (More than a year later, the army issues a final report concluding he was killed by friendly fire.) Marie asks how Christine's parents are doing. The families used to be neighbors in the same suburb, when the kids were teenagers.

"It's hard," Christine says. She isn't sure yet what her family will do for Christmas this year. It's better for her to spend Christmas with the Trykoskis.

"This war . . ." Jack says.

"Don't start," Marie warns her husband.

But Jack starts: Public approval for the White House's war in Iraq has taken a dive in 2006. U.S. military deaths in Iraq — including Bill Meeuwsen's — presently number about 2,900, just surpassing the death toll from the September 11, 2001, terrorist attacks in New York, Washington, and Pennsylvania. There will be 117 soldiers killed in Iraq this month alone. After a lengthy trial, Saddam Hussein is scheduled to hang (and does, two weeks from tonight). Meanwhile President George W. Bush will soon ask for an increase in troops for a maneuver soon to be known in the news as *the surge*.

Living in the bright red center of the culture that made the governor of Texas into a president, Jack has always been a Bush supporter. But he says he's changed his mind about how the war is going and whether the United States should continue fighting it this way. His work in the oil business has taken him many times to the Middle East. "I voted for him — twice," Jack says. "But they've done everything wrong. He won't listen. There are profound misunderstandings about [the culture] there."

Christine's eyes widen, then narrow. She clears her throat. She's not having it. "The media is distorting the picture," she says, and it hurts her to see Jack falling for it. "Good things are happening over there," she says. "We just aren't being told about it." The military effort has to succeed, and is succeeding, according to soldiers that she knows, friends of her brother.

Jack argues back: "That's not true. You —"

"No, it *is* true," says Christine. To pull troops back now will be the worst kind of defeat of all, and it will mean losing the cause for which Bill died.

The Trykoski Christmas has now shifted into the very debate the country is having everywhere. "Look, I'm very, very sorry about him," Jack says. "But the country can't go on like this. It's stupid."

"It's not stupid," Christine says. "It's a war."

"But you —"

"Jack, Jack, Jack," Marie says, pulling on her husband's shoulder, trying to get him out of the chair. "Let's go outside, I need a cigarette. Come on."

"Well, it's frustrating," says Jack. He can't have an opinion just because her brother died? He can't talk about things that he knows to be true, even if it hurts her feelings? "I've been there, I've been to these countries. It's not as simple as all this."

"Jack, we know, but drop it," Marie says. "Drop it."

Jack won't drop it, and Christine won't drop it. She's tough. She's been collecting boxes of goodies to send to AnySoldier .com. She believes the war can and will be won, as horrible as it may get.

Jeff and Bridgette? They both stare at the carpet. Doug gets in there a little bit, taking Christine's side. Traci fiddles with her bridal headpiece.

"You kids don't know," Jack says. "You're not old enough to know."

"Jack, come outside," Marie says. "Mr. Reporter Man," she says to me, "we are going off the record here, yes?"

My notebook closes.

A little while later, Greg follows me to my car. "Well, that was a good time, wasn't it?" he says. Bill's death hit him hard, too. We've talked about this before, one night in Greg's pickup, coming back from Hooters. Greg wonders if he shouldn't be over there with some of the friends who were in the Corps of Cadets with him at A&M, instead of working a job where, in that way that all things wind up seeming ironic if you think hard enough, he sells ExxonMobil's oil to manufacturers. He knows he needs to figure things out with Christine someday and think about their future. "Here, I got you a Christmas present," he then says, handing me a small envelope with a scantily clad chick wearing telltale orange hot pants on it: a Hooters gift card, $25.

He goes back in the house. A car pulls onto Bryson Drive, and the driver rolls down his window: "Are their lights done for the night?" he asks.

I tell him yes, their lights are done for the night.

Crèche

✦

(ONLY 5

SHOPPING DAYS

TILL CHRISTMAS!)

14

Things Remembered

T HE NATIONAL RETAIL FEDERATION, still loyal to its
initial September projection of a $457.4 billion holiday
season in overall sales, issues its usual baffled statement five
days before Christmas regarding the mental state of those who
have not yet bought a thing: 33 million Americans haven't
started their Christmas shopping this late in the game, and
most of them are men. "In ten years, this is the most last-minute
holiday season that we have ever seen," a Citigroup analyst,
Deborah Weinswig, tells ABC News.

There is always some new twist or permutation of holiday
economic dread for analysts to predict or caution against. This
time it's the specter of dismal sweater sales: the same year Al
Gore issues his environmental warnings in the documentary
An Inconvenient Truth, December temperatures have risen to
freakishly warm highs in much of the country, threatening the
very *feeling* of Christmas. Normally snowy, wintry regions of the
country are enjoying balmy days, meaning retailers are unable
to push holiday staples such as sweaters, coats, scarves, and
gloves. (It also means a drop in market demand for more sub-
liminal ideals: Charles Dickens, George Bailey, Hallmark movies, "Baby It's Cold Outside," and all that.) "The weather makes a

difference to me," one woman tells the *Dallas Morning News*, as she wanders Collin Creek Mall in Plano. "I can't shop for Christmas presents while I'm still wearing shorts." On December 23, the *New York Times* reports that retailers have dubbed this "the Coat Crisis of 2006," which "has sent a chill through the executive suites of major retailers" and has "merchants scanning weather forecasts" like rain-starved farmers. It appears as if every last item in the stores having to do with winter and coziness and warmth has been put on sale; the entire concept of bundling up and snuggling in is abandoned.

This veneer of blame, heard in a *Wall Street Journal*–ish, CNBC-ish tone, never changes in economic reports of Christmas commerce: It's your fault. You, the consumer. There is nothing more depressing about Christmas than to learn that all the shopping we've done is somehow always insufficient, even when it brings profits to shareholders. We are spending more than we spent last year, when we spent more than the year before that, which was more than the year before *that,* and still it is not good enough. Hundreds of billions of dollars and still a bummer, like having a person on our Christmas list we can never fully please, who acts out a key scene in any family's holiday dysfunction drama.

Economy: *You never get me what I want.*

Consumer: *What?! I buy you everything you ask for! Every year!*

Economy: *I still feel empty.*

About the mall, several Frisco residents tell me that although the decorations and bustle delight them as the season begins in November, it can also unnerve them as December 25 gets closer. They don't find the right gifts or start to feel guilty about the have-nots, and they sometimes feel Jesus watching. By the third week in December, shoppers take pleasure in informing one another (and me) that they have finished all their Christmas shopping, which means they no longer have to drive to, park at, or walk into any shopping mall anymore, ever again (although this is where I find them). They are drawn to the mall even as they find an elemental abhorrence in going.

One of my Frisco acquaintances, a man named Ben Beckelman, tells me his grand idea to bring balance and sanity to Christmas shopping: "I'd like sometime to get a guy who looks like Jesus, and have him carry a cross, flanked by two Roman centurions. You know? Just quietly carrying his cross through the mall. I think that would be kind of neat."

I imagine the route that the Stonebriar Stations of the Cross would take: Past the Aveda boutique, past the Build-A-Bear Workshop, past Things Remembered. Up the escalator, past the Abercrombie, toward the Gap. Stopping to greet the weeping women of StrollerFit. That's about as far as Christ with his cross and his soldier-tormentors would get before the Stonebriar Centre security guards would catch up to them, their walkie-talkies abuzz with secular alert.

I take a plane ride from a sightseeing outfit called Starlight Flight a few nights before Christmas. The owner, David Snell, says he does big business in December because people want to see the holiday lights in an all-encompassing way. He sells it as a romantic evening out; men sometimes propose to their girlfriends on these trips. It costs $235 for a one-hour flight for two passengers.

I sit in the front with a young pilot who works for Snell; in the back sit a man and a woman who are on a date. We take off from a business-jet airport in Addison, near Plano, off the Dallas North Tollway. There's a strong, cold wind blowing across the runway, and the single-prop Cessna buzzes loudly as we fly over glass-paneled office buildings and the string of red brake

213

lights of tollway traffic, upward and out over the illuminated grid of 6 million people.

The pilot keeps pointing out Christmas lights, which are nice, but I keep looking at the mall parking lots. We ascend over the tollway, over and past the fabulous Galleria mall at the LBJ Freeway, then over Dallas's toniest neighborhoods, then out toward the town of Arlington and its Six Flags Over Texas amusement park; we then circle back, south of the neon-limned buildings of downtown Dallas, fly along U.S. 75 through Plano, and finally bank west to Frisco. There's Jeff and Bridgette Trykoski's house, the easiest to spot. We make several loops over it, and I wonder if they're home. We turn left again and soon below there's Main Street, Frisco Square, and City Hall, with all of Jeff's 150,000 lights blinking frenetically to a soundless Mariah Carey.

A few miles south now, we fly over Stonebriar Centre mall and the Centre at Preston Ridge strip mall's big-box stores. The sodium-vapor lamps in the parking lots cast perfect circles of white and orange light. Even with the turbulence that has dogged us the whole way, I am filled with serenity, floating above. This is it. The tiny people in their tiny-big houses and tiny-big malls and tiny-big churches, making Christmas happen as best they can. Below me is that perfect, porcelain Department 56 village, all arranged. Here is that dream the mayor had, the one about flying over new developments. The godly omniscience up here. The wind tossing our Santa sleigh up here. The grid sprawling toward forever, twinkling into the horizon in all directions, all this land that people keep saying used to be "nothing."

When I got here.

Where the Home Depot is now.

Where there were just cows.

* * *

214

The present occasionally overwhelms me, as it would anyone
— the traffic, the text messages, trying to decide whether to eat
at Applebee's or Red Lobster. On days when I've had enough of
malls and megachurches, I can always find solace in such things
as Collin County records from the late nineteenth century, or
in the maps to overgrown cemetery plots, or deep within reels
of newspaper microfilm. During my first weeks in Frisco, I lost
two whole days sitting in a library and reading about the dirt of
North Texas — soil layers, shale striations — for no real purpose
other than it seemed the reason why anyone ever came here in
the first place: the ground itself. The free land.

Going a hundred years back now; cranking the microfilm by
hand, watching pages stream further into Christmases past,
when there was barely a Frisco there to be found on the map,
when the railroad had only just opened. It seems barren and
hard, until you encounter the wistful schoolgirl poems about
cowboys and willow trees, and little church news items that in-
dicate a paradise in the rough. Aside from census data, there's
very little in the way of a reliable, detailed record of the few
hundred people who lived here at the turn of the twentieth century, other than those who owned the land and those who could
afford tombstones.

It sounds a lot longer ago than it was. In some ways it seems
barely a minute. With Dallas a train ride away, the nearest "big"
town to Frisco then was McKinney, the Collin County seat, seventeen miles east with a population then of 4,342. Searching
through microfilm one afternoon to see what Christmas sales
the local merchants advertised back then, I come across a feature that ran in McKinney's *Daily Courier* newspaper in December 1902 called "Little Folks' Letters to Santa Claus":

"Dear Old Santa Claus," wrote one boy on December 22 that
year. "As the time is drawing near for you to come, I will tell you
what I want you to bring me. Old Santa, please bring me a drum,

foot ball, Roman candles, fire crackers, candy, apples, oranges and nuts. Bring my brother Wiley, Fred Burnitt and my little baby brother something too. Your little Friend, Avery Bearden."

"Dear Santa Claus," wrote a girl named Esther Brown. "Bring me a doll, a dresser, a book and a side board. I love you a hundred bushels."

"Dear Old Santa: I want you to bring me a doll and a piano and any thing else you can spare that is nice for little girls. Yours very truly, Willie C. Phillips."

At the time these letters were written, the icon of Santa Claus (the one we know — jolly and fat, arriving by airborne sleigh, coming down the chimney) had been fixed in the American imagination for only eighty years or so, particularly since the Manhattan land baron Clement Clarke Moore wrote (or is credited with writing) *A Visit from St. Nicholas* (aka *'Twas the Night Before Christmas*) in 1823. Santa had a remarkably rapid ascension to celebrity status across a largely pioneer continent without the help of televised cartoon specials. I jump ahead one microfilm reel to the *Daily Courier* issues of December 1903 and notice that the tenor of the children's letters seems to ramp up, becoming more terse and specific, especially now that the children have the *Courier* acting as a public forum where they can express their desire and deserving piety: "Dear Santa Claus: As you have always brought me something, this time I want you to bring me a nice story book, a suit of close [*sic*] and cap and a pocket book. Joe Largent."

"Dear Santa Claus: Please bring me a doll carriage and a doll house, a story book and a toy telephone or anything that you have to spare. I don't want to be greedy. Your little girl, Lelia Foster."

"Dear Santa Claus: I want a doll. I want a baby buggy and a doll and a little toy wagon and some candy and oranges and some nuts. Please dear Santa Claus take some little poor girl a doll and things. I want a little horse to hitch up to my wagon. I like picture books. Your loving little girl, Olive Elizabeth Bush."

"Dear Santa Claus," began a letter the *Courier* published on Christmas Eve, 1903, a dispatch that seems to be written in an imitation of toddler-speak:

"I hope I have not waited too long before I wrote you dis 'ittle letter, telling you what I want 'oo to bing me Tistmas; you tan find my 'ittle 'tocking besides titter's on the 'ittle chair, I want a 'ittle doll bed, some candy, oranges, apples, and some of those good nuts called negro toes. Bing titter Edna Mae a 'ittle doll too. Don't forget my school teacher and bing her a new hat. I close with my love to you and I hope your reindeers are well. Your little girl, Hattie Venus."

These are America's favorite Christmas children, emerging from the faraway influence of the last days of the Victorian and pioneer age, captured in letters to the jolly old elf. They are contemporaries of Virginia O'Hanlon of "Yes, Virginia" fame. Their desires seem almost ascetic compared with our own; they want nothing more than dolls, wagons, a "suit of close." They ask only for apples and oranges and, unfortunately, "negro toes" (Brazil nuts). People still dream of such cherubs and portray them on Christmas cards and in advertising. Frozen in time, they represent an antithesis to Wal-Mart, to Bratz (and brats), a nostalgic relief from our rapacious worst. Innocent of high technology and plastic, they awaited a "Christmas season" that came and went in a matter of days.

McKinney and Frisco now compete for newer and better retail developments and custom-made show-home neighborhoods, swiftly developing the land into pseudo-villages. The Dallas North Tollway construction in Frisco is stopped for three days in September 2006 to unearth the forgotten graves of an infant girl and a teenage boy, buried there in 1902 and 1905, it is believed. The bones are taken away for possible DNA analysis; there are questions of whose remains they are, since the private family cemetery that had been here was reportedly exhumed and the occupants reinterred at a cemetery in Frisco in the 1970s. The old graveyard is directly atop the site for a tollway underpass.

It is believed by some that the baby girl died on December 14, 1902, and that she belonged to another family on whom the original landowners may have taken pity and offered a plot. She was buried in a small wooden box hammered together with large nails. Between the "Little Folks' Letters to Santa Claus" and the mercantile Christmas ads, there's not a word in the old issues of the *Daily Courier* about such a death. A stillborn would have been a routine tragedy and, given the distance to Frisco, too far away to note. The meteorological records show a half-inch of precipitation with temperatures in the high thirties on that December 14 — a sleety, easily morose day. It would have been difficult to dig a grave in the blackish gray soil.

After the North Texas Tollway Authority's archaeologists unearth the baby's grave, there is a photograph on the front of the Metro section of the *Dallas Morning News*. Three men in safety vests, hardhats, and sunglasses are hunched over a hole. Next to them is a man sitting on an overturned bucket, his hands clasped in front of him, almost as if in prayer. Behind them, in the distance past orange-and-white barricades, is a 7-Eleven on the frontage road. Everyone in the picture is looking at the hole in the ground. I clip this picture out and tack it on my wall.

Another discovery is made a day later: "2nd Grave Found at Tollway Site" reports the *Morning News*. These are the bones of a young man or teenage boy, buried a few feet from the first grave, contained within the rotted remains of a more elaborate casket. I go out to the construction site the next afternoon, a Saturday, with a Big Gulp in one hand and my notebook in the other and my curiosity about what was once here and what is now gone. Standing with his arms folded, the chief engineer assures me the archaeologists have declared the dig over, even with the possibility that two other graves had once been here, too. The bulldozers will start up again Monday to scrape out a twenty-foot trench for the new tollway lanes.

Weeks later, nearer to Christmas, it hits me that the picture I'd clipped from the paper of the gravesite dig is composed like a manger scene, a crèche. Everyone in it is looking in wonder at the spot where the mystery baby lay.

Going forward now, Christmas present. On a Thursday night in December, I reconnect with Caroll Cavazos and her daughter Marissa for dinner at a Souper Salad in Carrollton, before we go to Marissa's school, American Heritage Academy, for the fourth grade's Christmas play.

Over her soup, Marissa tells us the Nativity story, her way: "Mary put her hands on her belly and yelled 'Fire in the hole!' and out popped Baby Jesus."

"Marissa, *stop*," Caroll says. "Where did you hear that?"

"My school is off the heezy," Marissa tells me proudly. It's another of the area's many private religious schools, requiring as a condition of admission that at least one parent of the pupil profess a born-again belief in Jesus's saving grace and costing about $6,000 a year in tuition and fees (Caroll gets a discount for being a tithing servant-leader at Celebration Covenant, which has ties to the church that operates American Heritage; the school is a relative bargain compared with similar schools that cost upward of $15,000 a year). Caroll says private school is among her greatest extravagances and worth it, "for teaching the values that I share."

Tonight the fourth grade is presenting a copyright-*in-flagrante* production of *A Charlie Brown Christmas*. It's word-for-word loyal to the classic 1965 television special, down to a Snoopy doghouse and Lucy's 5-cent psychiatry stand. When we get to the school, Marissa runs off to get in costume, and Caroll and I find seats in the packed cafeteria-auditorium. "I get butterflies," Caroll says. "I know this should be no big deal for her, but I'm always nervous before she goes on."

The lights dim, and the audience hushes, and a jazz combo (the music director and two other men) plays those first familiar notes, *Christmas time is here / Happiness and cheer* . . . Marissa is playing the plum role of Lucy van Pelt. More correctly, she is one of three girls alternating in the part from scene to scene, appearing in the middle scenes, where prima-donna Lucy threatens Linus with "five good reasons" to give up his blue blanket — "One, two, three, four, five," she growls, folding her fingers into a fist, louder and stronger than almost any of her cast mates.

"Overacting" is usually Caroll's only note to Marissa after a stage performance or audition: too much of the wrong kind of energy, take it down a notch. It is also Caroll's ongoing request to her daughter in everyday life, when Marissa interrupts Caroll in the middle of a conversation with someone else, or sings to the cat, or pirouettes back and forth across the kitchen linoleum in her older sister's platform shoes: "Marissa, calm down."

The fourth graders (with the third graders singing in a chorus) achieve that perfect melancholy blueness and peacefulness of Charles Schulz's lasting intent. During the last round of "Hark, the Herald Angels Sing," Caroll gets a little teary. Everyone does. After the curtain call, the director is presented with an engorged gift basket and a floral arrangement.

Caroll and Marissa spend a considerable amount of time pursuing Marissa's dream of becoming an actress. Caroll has waited

through enough of Marissa's weekend classes at a children's acting school to absorb some of the key lessons, so she is always encouraging Marissa to be more *real*. In addition to acting classes,
Caroll also paid for moviemaking summer camp, which ended
with the "red-carpet premiere," at a local cinema-restaurant in
Plano, of a short digital film that everyone in the family (even
Marissa) pronounced unwatchable. Caroll has paid for a series
of head shots of Marissa in various guises: little girl in braids,
green-eyed glamour princess, bookish nerd, spunky tween. Caroll has also sent Marissa to a six-session acting series taught by
the mother of a kid who's played one of Hannah Montana's boy-
friends on the Disney Channel.

Marissa has met with agents and casting directors — all the
acting class students do, eventually, when such California digni-
taries visit Dallas, which causes great excitement if not actual
bookings. Caroll has sometimes wondered, while looking up
other children's acting bios on IMDB.com, if God isn't trying to
lure her and her daughter to Hollywood. Once in a while, one of
the kids in acting class gets a big break, and soon enough Caroll
will be hearing of booked commercials, or Disney sitcoms, or
parts in feature films. The more aggressive kids and parents
have been known to head to L.A. in January, to audition for
TV roles during what's known as pilot season. Marissa hears the
words *pilot season* spoken by grownups with the same dreamy
regard with which Pastor Keith talks about heaven.

"I tell Marissa, if that's God's plan for her, then it's our job to
be ready to listen for it," Caroll says. "I may be wrong, but I really believe that there is going to be a rise in Christian entertainment, and not the cheap stuff, like *VeggieTales,* but wholesome
stuff. Pastor Keith has said that church should be the most creative place, as creative as anything in Hollywood, and I think it's
coming."

* * *

Caroll saw a documentary on TV a few days ago, on the USA Network, where a camera crew followed a few couples, families, and single people through their Christmas plans. The single people found love, and the married people seemed to fall more in love, all thanks to Christmas. It was a nice show with soft edges, and although she enjoyed it, Caroll says it was a little hard to take. This is the first Christmas in a long time when she realized something: she's been feeling lonely. "That's a weird feeling, because I'm never really alone," she says, since Marissa is always around, and Ryan has moved back home. She'll be fifty in a few months. In addition to keeping her eyes and ears open for the moment God may want Marissa on the Disney Channel, she's trying to keep herself open to one last chance at love.

It's a Sunday night in mid-December and Marissa goes with Caroll to the Christmas party for the church's singles ministry. (I come along, too.) It's a potluck dinner at someone's house in the northwest end of Frisco, near the construction site of a gargantuan new Wal-Mart. One of the bedrooms is set up for kids, with a video playing Christmas cartoons, which Marissa quickly pronounces boring, affixing herself to Caroll's side.

The main event at the party is another mystery-gift swap like the one Tammie Parnell's Hotties had done, but this time with four dozen participants, and with much more squealing, as everyone tries to steal candles, Starbucks gift cards, DVDs, and inspirational ceramic crosses from one another. Out of a red Santa hat Caroll has drawn number 1, which is just the sort of thing that terrifies her — going first. The room watches while she unwraps a cheap chocolate fondue pot about the size of a large coffee mug. Marissa gasps: "Mommy, I've always wanted one of these!!" (Nothing can be done about foisting the fondue pot onto someone else, but when my number is drawn I try to improve Caroll's night by nabbing a soundtrack CD of *A Charlie Brown Christmas*, which Caroll clearly wishes she'd picked, from a plump blond woman who seems none too pleased with my aggressive play.)

The party ends with Associate Pastor Ray Harmon leading a round-robin prayer. We all hold hands and ask God to open our eyes beyond the materialism of the season. A roomful of pretty single women in sparkly blouses and full makeup nod in thoughtful assent, as do the dozen or so single men in their best golf shirts. One woman tells the group that her recent divorce had turned her "Scroogey" this year, until she started buying presents for the Angel Tree at work. Associate Pastor Ray mentions how, the other day on *Oprah*, everyone in the audience got a thousand bucks on a credit card, but then Oprah told them they could spend it only on a charitable cause. ("With Bank of America credit cards," Caroll notes, proudly, of her employer.) One man tells the group that this Christmas he's been trying to give up buying things with credit cards, to get past "meaningless debt." But another man responds that a credit card is okay, and that he has decided to "give over my American Express Blue to God, which I'm using to tithe on." He now trusts the Lord to provide for the balance.

Caroll and I are wandering through the holiday hordes at Stonebriar on a Saturday afternoon, a week after *Charlie Brown* and the singles party. Marissa darts into American Eagle to look around. It is here that Caroll tells me Marissa is her best friend. Wherever Caroll goes, Marissa goes, too. At a Celebration Covenant Church function the previous spring, Pastor Sheila Craft had told the crowd that Caroll was an extraordinary servant-leader and called her up to the stage, in front of everyone, and surprised her with a big gift: a free Caribbean cruise for two.

But it came with the condition that Caroll could not take Marissa with her. There was an implied message — in front of the crowd — that Caroll needed to do something just for Caroll, to have a life. Caroll searched for another adult to ask along on the cruise and, after much thought, finally asked another single woman she knew from church. As the departure date approached, the woman came down with a rare disease and backed out, so Caroll decided not to go. Now, months later, the friend has moved away and has so far failed to produce the right medical paperwork so Caroll can get a refund or book another trip.

Caroll considers this chain of events to be somewhat typical. She senses the approach of small disappointments here and there and prays every day that God will take her past that thinking. Pastor Keith has said many times, many ways, that we must "Think and Be and Do transformation." She is still not sure what a transformed Caroll looks like. The stage note Caroll usually gives to herself is *more energy, be stronger.*

Caroll usually puts off most of her Christmas decorating as a way of honoring her son Ryan's birthday on December 16. She thinks it's a bummer if your birthday gets swept away by the Lord's, so she always waits to unbox, assemble, and decorate the tree until after the seventeenth.

Ryan's turning twenty. The plan has been for Caroll, Marissa, and Ryan to meet up with Caroll's older daughter, Michelle, and Michelle's husband, Joey, for steak dinners at Texas Roadhouse. But Ryan forgot he'd switched shifts with a coworker, so he has to work tonight at Best Buy.

Which is where I almost always find him. He is back in the "car-fi" department, in his low-slung khakis and Best Buy blue shirt. He is talkative and always friendly. He decided to keep one of the $600 PlayStation 3 consoles that last month he and his friends had camped out in shifts to be the first to buy at Circuit City, so they could resell them on eBay at a profit. "That plan pretty much fell apart," he says. Now it's all about the Nintendo Wii, which just came out and, Ryan accurately predicts, "is going to be huge." He points toward the store's lone display model of a Wii. "We're getting a shipment on Sunday. I might try to get one."

He's been working at Best Buy for six months. In the run-up to Christmas, he says his department has been pulling down daily sales totals that top $23,000. (That will drop below $5,000 a day once Christmas is over, which is when employee shifts will also be reduced.) They sell a lot of satellite radios. People are also starting to want global positioning devices for their cars. "GPS — that's what everybody will be asking for next Christmas," he says. (He's right. Two years from now, most of Best Buy's car-fi department will become "mobile electronics," mainly for GPS, and Ryan will still be working there.)

With a GPS, you always know where to turn. The guessing is gone, but the price for that, psychologically, is this sense of never going astray but never quite knowing where you are, instinctively, among all this sameness. The consumer species can no longer abide unknown outcomes. Most new cars come now with an option for emergency buttons to press in any sort of mishap, so a live operator will come on and tell you what to do next, in a reassuring voice. The GPS knows where all the Best Buys are, where the Chick-fil-As are, where the new tollway on-ramps are, and which exits to take to get off. So far, Ryan hasn't installed one in his used Ford Explorer, to go with his massive sound system. He likes the idea, though: never being lost, always knowing the surest way.

I leave Best Buy realizing I am likely to be among the last of the American meanderers, willing to drift off the map. Here there are no mountains, no wide rivers, or contours, or valleys. The neighborhoods all have market-tested names that feature the words *Preston* or *Stone* used in infinite combination with often irrelevant geographical traits: *Creek* (no creek), *Lake* (man-made collection ponds) or *Lakes* (same), *Vistas* (none), *Ridges* (barely), *Bluffs* (not), and *The Preserve* (of?). Often, people tell you where they live, and how to get there, by their distance from stores.

We are two lights past the Home Depot.

If you see the Taco Bueno, then you've gone too far.

One morning I see a heron take flight, regally, from the construction site of a new Holiday Inn business hotel. It zooms low toward the back end of Linens 'n Things. Another morning, near Ryan's Best Buy, a coyote pup dashes between Pier 1 and the Ross Dress for Less as I cut across the parking lot to save time. I blink and barely glimpse him.

15

The Pageant

E ITAN THE ISRAELI has disappeared from the Snow Powder kiosk in Stonebriar Centre, along with his girlfriend, Tali. It's as if they never existed. Now there is David, the new Israeli, left here with too much surplus Snow Powder, saying he's never heard of Eitan and Tali — he just got here from another mall.

The Gap employees have given up on folding the sweaters, which now create a twisted, cashmere-blend and merino-wool knoll on a counter in the center of the store. This is the moment Christmas goes into active labor. With only three shopping days left, the mall takes on that full-to-bursting feeling of release, expunge, surrender — the push before the big push. The manger baby is coming.

In the mall concourse, we straggling, last-minute consumers are serenaded by a trio of Andean pan flutists from South America playing versions of "Away in a Manger" and "God Rest Ye Merry Gentlemen," sounding like Christmas music you'd hear at a massage spa in New Mexico. I walk upstairs to the AMC 24 multiplex and buy a ticket to the 4:40 matinee of *The Nativity Story*.

This movie is supposed to do for the crèche what Mel Gibson's *The Passion of the Christ* did for the Stations of the Cross

($604 million worldwide box office, a major hit), yet, as those pesky godless movie critics have pointed out, *The Nativity Story* turns out to be beautifully uninspired, and the churchy audiences that propelled the success of *Passion* have curiously not been filling the nation's multiplexes. Since its opening four weeks before Christmas, the movie has taken in $33 million, or $2 million less than it cost to make. Bigger hits this Christmas are *Deck the Halls*, in which Danny DeVito and Matthew Broderick stage a resentful war of Christmas lights ($34 million), and *The Holiday*, in which Cameron Diaz and Kate Winslet swap houses and Christmastime romance woes ($36 million in two weeks, so far). Everyone else went to the movie about the ostracized penguin that dances (*Happy Feet*, $160 million to date).

But *Variety*'s number crunchers have no case to make here at the Stonebriar AMC 24. The only seats left in this packed showing of *The Nativity Story* are down near the front, and I settle in just as it begins: Virginal Mary says yes to the angel. A scandalized (adorable) Joseph finally accepts that her pregnancy is a gift from God. A census order goes out across the land. There are pretty panoramic views of Roman-era Palestinian sunsets and icy purple skies. Herod is evil and wears a lot of fey mascara. The Magi are seen as comic relief. Labor pains start. No room in the inn. The manger. The hay. The animals. The white light of the star of Bethlehem illuminates the theater and I turn in my seat and admire the audience — adults and kids, their faces aglow in slackened satisfaction. The moment: Joseph holds Mary's baby aloft.

I notice an improvement on the miracle: there is no umbilical cord.

Here he is, everyone.

Jesus, unplugged.

* * *

The Christmas pageant at Celebration Covenant Church is scheduled for three performances, which will take the place of regular weekend services — one on Saturday night, December 23, and back-to-back shows Sunday morning, Christmas Eve.

Caroll, the dutiful servant-leader, has volunteered her time as a technical assistant, and she's gone to all the rehearsals; Marissa, having absorbed the first rule of acting class (no small parts, only small actors), has accepted her brief stage time in a Victorian crowd and snowball fight scene.

Two days before dress rehearsal Caroll, Marissa, and Ryan drove three hours up to Moore, Oklahoma, to spend a night at Caroll's mother's house. Caroll tossed and turned on the guest bed, unable to sleep, and Marissa seemed to be coming down with a cold the minute they arrived.

Caroll is fond of her mother and talks to her daily on the phone. It's not that she dislikes going to Oklahoma; it's just that the memories are not always good. Stay away from home long enough and you start to realize why you had to leave. Then Christmas comes around and you plan for the most efficient trip back, with that odd pull of duty and love. Caroll didn't see her father after about 1970, when he left for a tour of duty in Vietnam and left Caroll's mother and the children, too. As an adult, Caroll says she recovered early childhood memories of sexual abuse by her father (who has since died). She credits a therapist for working her through it. "She grew me up real good," Caroll says.

That stuff is so far back it may as well be like the memory of that shadow of the rocking horse on the wall, the shadow she saw on a Christmas morning when she was three or four. It hasn't been an easy life since she left Oklahoma — divorces, single motherhood — but Caroll would be the first to tell you that faith will find a way to see you through. Her confidence comes and goes, but by God, she's forged a career, paid for a house, and kept herself current, fit, and together. *And her Christmas shopping is done*. Once in a while she allows herself the most minuscule amount of pride.

Caroll tried to keep things light with her mother, who still wants a mink coat instead of the computer Caroll bought for her on Black Friday. "Oh, well," Caroll told her, and soon enough the computer was connected and her mother was online. Caroll and the kids are back in Texas a day later, in time for Ryan's late shift at Best Buy.

After dropping him off, Caroll and Marissa go directly to Celebration Covenant for the pageant dress rehearsal. Caroll puts on her headset and has a look at the stage. "I guess the show is pretty messed up right now, but they always pull it together," she says.

Gina Anderson, the show's director, has the intense air of the theatrical profession about her, checking a clipboard and issuing commands to cast members, singers, and crew. More than a hundred church members have volunteered for the Nativity pageant, which is being called *It's a Wonderful Life 2*. A series of musical numbers will bookend a simple story (written by another church member) of a Victorian Age businessman-type husband and father who realizes he's not spending enough time praying to God and loving his family.

The church's main stage has been snow-villaged out as a turn-of-the-century town square, covered in Polyform snowdrifts, bare trees, old-fashioned gas lamps, park benches, and a genuine Model T parked stage left. The scrims are backlit in a wintry lavender. At stage left there's an old-fashioned living room with a settee, a piano, and a Christmas tree. Since the music is prerecorded, the stage doesn't need to accommodate Celebration's large praise band; the music director, a folksy, middle-aged man with a trendy stubble goatee and frosted blond tips in his spiky hair, will play a lead role as a sage and songful family friend, a Clarence-the-angel type. The forty or so people in the choir and cast have all been outfitted in Victorian garb — the men in top hats and tailcoats, and the women in long bustle skirts and plumage hats.

Gina and Caroll take a seat in the front row. The cue is given for the dress rehearsal to begin. Since Celebration Covenant does not lack for star quality, or the dream that anyone could be the next Carrie Underwood or Chris Daughtry, all the singers are fully in touch with their inner *American Idol*. The show opens with a hyper-happy medley, as the townsfolk sing "It's the Most Wonderful Time of the Year," while Marissa and the other children throw Styrofoam snowballs at one another. This is followed by "Frosty the Snowman" (a shaggy-haired teenage boy has that role, wearing a tattered, rented snowman costume and giant head). Here, several little girls — some dressed as wrapped presents and Christmas trees, and some dressed as penguins in yellow swimming flippers — waddle onto the stage for a dance, then waddle off when the number ends and the lights go down. The lights come back up on the Victorian living room drama. As the main character listens to his preacher/angel friend and searches his soul, Pastor Keith and Pastor Sheila's teenage daughter, Whitney, performs her big, melismatic solo number ("Where Are You, Christmas?"), failing to locate some of the notes. Then there's a ballet, and then there's "Away in a Manger," sung by a few dozen children in matching red turtlenecks. Then there's a Nativity, with shepherds, Joseph, Mary, and a real infant as Baby J. (A doll understudies tonight.)

The final number brings everyone back on stage: all the Victorian chorus members along with the little girls as presents and penguins, the red-turtleneck children, the ballerinas, the Holy Family, Frosty the Snowman, and, from stage right, an adult Christ stripped down to his loincloth and smeared with Dracula blood, dragging a cross to center stage while being whipped by two centurion guards. (I recognize the bloody Christ from the singles party.) Here is where the Nativity, Dickens, and Burl Ives collide head-on with Good Friday, as Jesus is crucified while everyone sings "Hark, the Herald Angels Sing," ending on a long, noisy note: "newborn kiiiiiiiiiiiiiiiiiiing."

Then they freeze.

Hold it for applause.

Here is Christmas reinterpreted as Picasso's *Guernica,* everyone crammed together, permanently caught in a nightmare.

Director Gina is not pleased, yet she tries to smile as she reviews her many notes for the cast and crew. Nearly everyone missed his or her cues. Nobody got on or off stage in time. Lyrics were forgotten or mangled. It took them more than an hour to get through a show that is supposed to be forty minutes long. "Let's do it again," Gina says.

Caroll's job is to communicate on her headset with the stage managers and the guys in the lighting and sound booth, watching for missed cues and other glitches. On the drive back from Oklahoma this afternoon, Caroll let Marissa have some Sudafed but forgot to check if it causes drowsiness. Now it's past 9 P.M., and Marissa is whiny and curling up next to Caroll in the front row, asking how soon they can leave. The lights go down and the cast takes it from the top. The kids in the snowball fight are supposed to be backstage, getting ready to come on, but Marissa, still in costume, declines to budge from her mother's side. "Mommy, I don't feel good," she says.

"Marissa, I know you're tired," Caroll says. "But I can't have you on me like this. Sit in your own chair. Lay down if you want to."

A woman in a bonnet sings her solo of the second verse of "It's the Most Wonderful Time of the Year," too loud and showy, half an octave off, and Caroll looks at me and rolls her eyes with the first measure of church sarcasm I've seen (or will ever see) from her. Then comes the waltzing scene and the snowball fight, and pretty soon Frosty comes on, then the little girls —

"They're not moving all the way down to the right side," Gina says aloud, pointing to a group of extras stage right. "There's not enough room for the penguins. Caroll, can you go up there and get them to move over? They need —"

Caroll is up and on it.

She climbs the five steps to center stage, keeping her head low. She hates being on stage. (If she liked it, she'd be in a long dress and bustle right now, waltzing back and forth and singing "Winter Wonderland.") She tells the stagehand behind the curtain that the singers are too bunched up on this side, that Frosty can't get through — and here comes the Frosty song, and Caroll bends low and returns to the edge of the stage. The lights go dark, because there's a scripted bit where one of the snowball children "trips" over an electrical cord socket and says, "Oops!" and plugs it back in, and "Frosty the Snowman" starts up again. In that brief darkness, Caroll tumbles roughly off the stage, falling the remaining three feet down the steps. She rolls to the floor and winces in pain and yells out, but no one can hear her. The penguins and the dancing presents are on stage now, the singers singing, "I'll be back again somedayyyy!"

"Oh, oh, owwww, dang it!" Caroll yells.

Gina looks over to see Caroll writhing and clutching her left foot. "I can't, I can't," Caroll says, holding back tears. Which foot? "Both of them," she says, not sure.

Now Marissa has noticed, sitting up in her chair, blinking.

As the show blares on, Gina tells another cast member to go get a man playing one of the Victorian singers. He's a doctor. Gina lays hands on Caroll's leg: "Jesus," Gina prays aloud, closing her eyes. "Jesus, we know you're watching us, we know you are with us, and Lord, we believe that you are going to take care of Caroll, that she is going to get up and walk right now, Lord, that you are working through us right here, right now, and she is going to be just fine, Lord. She is going to get up and walk, Jesus. We pray in Jesus's name. You are here. You are here and you are taking care of her!"

Gina takes her hands off Caroll's left leg.

Now is Caroll's cue to get up and walk it off.

She can't. She winces and sucks in a deep breath. All she can think is that she's somehow broken bones in both feet. I suggest taking Caroll to the hospital, because I'm sure I heard a bone crack when she fell. "I heard it, too," Caroll says. "Oh, I'm so stupid. I just missed that third step. I can't believe I've done this to myself. Oh, no, oh, no."

The doctor/singer arrives from stage left, wearing a top hat and tails. While the show continues, he and I carry Caroll up the aisle and out to the church lobby. We drag two overstuffed chairs from the office reception area, set Caroll on one, and prop her feet on the other. Caroll fights back tears. The doctor helps her get her shoes and socks off. A woman watching this says, "It's always a day like this when you didn't have your toenails done, huh?" and then wanders off. It's clear that Caroll's right foot is starting to swell. "Here?" the doctor says, pressing. "Here?"

"Oooh, there! Ow," Caroll says.

"Let's go to the hospital," I say. "I'll get my car. Marissa can come with us."

"No, I think I can do it," Caroll says, trying to set her foot on the floor. "Owww!" She falls back in the chair. The doctor then has to leave, to make his cue on stage. Ballerinas and shepherds wander by in the lobby, oblivious to her or uninterested. One shepherd, barely out of his teens, walks up to us saying,

"Owwww," and Caroll thinks he's talking about her: "I know," she says, "I fell off the stage."

"No, it's this," he says, of the rope holding his shepherd blanket in place on his head. "It hurts." Then he walks away. Another Victorian lady comes up and asks what happened, but before Caroll or I can tell her, she is telling us a long story in her Texas twang about her knee surgery last year. Another Victorian woman comes by and does the same exact thing — asks what happened, doesn't wait for the answer, but tells us about the time she broke her ankle and what it felt like and how long it took to heal. A servant-leader walks by and wonders if we should be using these chairs to keep Caroll's leg propped up, since they belong in the waiting area outside Pastor Keith's office. I begin to feel like I'm watching some Gospel parable unfold: a woman falls in a church and hurts herself badly enough to need to be carried out, and, one by one, her fellow devoted believers pass by to tell us their own personal stories of sprained ankles, broken legs, and various ailments.

As the second rehearsal is ending, Caroll gets up and hobbles out to her car with Marissa. When they get home, she pops Tylenol and asks God once more to make it better. On Friday, it seems the left foot is just sore, but now she can barely stand on the right. She stays off it until she can't take the pain anymore and drives to a big strip mall, to see, as she calls it, a "doc-in-a-box," who x-rays it and determines that there are no bone fractures or torn ligaments. "I almost couldn't believe it," Caroll says. "I heard it crack, you heard it crack, but he said it's just going to be real sore. He told me to stay off it." I suggest that maybe now Caroll can just skip the Christmas pageant altogether. She already has the day off work. But she is determined to servant-lead at the final dress rehearsal, the Saturday night performance, and the two performances on Sunday morning. Caroll is back at church that night, wearing on her right foot a support boot the doctor gave her. She puts her headset on and limps through another Nativity.

Taking my seat Saturday night at the Christmas pageant, I'm expecting to see the worst. That's what a guy like me does. (A guy like me also goes running off to hospitals when he thinks he's broken his foot. Jesus can stay for rehearsal if he wants.)

But *It's a Wonderful Life 2* somehow comes off crisp and clean. It is delicious holiday cheese. There are barely any flubbed cues or mangled notes, and it clocks in at precisely forty-two minutes, getting better with the two Sunday performances, when it's standing room only in Celebration Covenant's thousand-seat sanctuary. The church is filled with families dressed in new sweaters and faux-fur-trimmed shawls, a velvet bow on the head of every plaid-skirted little girl. The penguin dancers are a hit; the children's choir is camcorded by parents from every possible angle. Baby Jesus never cries, and grown-up Jesus seems to have applied some bronzer between Thursday's rehearsal and tonight.

After the finale, Pastor Keith comes on stage and sits on a stool to serenade his flock with his own solo ballad, a gospel Christmas hit from the 1970s called "This Little Child." He is bathed in blue light, wearing a many-buttoned suit and a cashmere scarf.

"Isn't God good?" Pastor Keith asks the audience, when he finishes his song, and the audience coos its satisfied *amen*s.

"I'm so glad you're here tonight," he says, noting all the visitors in the room. "Let us believe for you. Mary didn't know what to believe. Joseph didn't know what to believe. But they believed just enough to receive the Word Become Flesh . . . So just say to God, 'I'm going to believe you really did get a virgin pregnant. I'm really going to believe that.' That's a hard one to swallow. [But] say, 'I'm going to believe this year that you really are God and give you the opportunity to show your God-ness in my life . . .'"

Pastor Keith works himself up: "If anyone will open that door, God is saying, 'I will come in! I will come in and I will take up residence in your heart! And I will be God. With. You.' . . . And if you can't do it, let us believe for you. What a privilege, on Christmas, that we have. What a privilege to believe. Let us do it for you if you're not ready to do it."

Pastor Sheila comes up on stage and talks about how eager they are to open their Christmas presents at the Craft house. She says she's dying for some jewelry. Pastor Keith's mother comes up and leads us in another prayer. When we all stand, my hands are down and Caroll's hands are up, believing for me.

After the third and final performance on Sunday, Caroll limps from the car to her house. She tells Marissa, "Let's just get our pajamas on and watch TV the rest of the day."

It's the middle of the afternoon on Christmas Eve.

It's too late to make Christmas perfect, she decides. Little things slipped by. There are the lights she never could get Ryan to help her hang on the house. She wanted to hook up the manger scene figures for the lawn, but that never happened either.

Caroll takes some painkillers the doctor gave her and turns on "The 25 Days of Christmas," a marathon of made-for-TV movies and holiday specials on ABC Family. There's *Frosty's Christmas in July*, and *The Story of the First Christmas Snow*, and *Nestor, the Long-Eared Christmas Donkey*. There's *'Twas the Night Before Christmas*, and *Santa Claus Is Comin' to Town*, then *The Year Without Santa Claus*. All our stop-motion Plasticine friends are here.

Caroll gets on the couch, puts her throbbing feet up, and lets Christmas come to her in a fifty-inch flat-screen surround-sound fog. Here, on television, it is always the best Christmas possible. Broken hearts are mended just when it begins to snow. Families gather around the tree. And the warmest part: the commercials. To the tune of "Winter Wonderland," MasterCard is giving away a house ("Cheese and cracker platter: $30. Big roast turkey and all the trimmings: $55. Warm baked apple pie: $12. A full house: *priceless* . . .") and you get a chance to win it every time you use your MasterCard. There is Kmart, lovely Kmart, "where Christmas comes together." Elves are pushing carts through the BigLots! store. The Aflac insurance duck is stuck in the chimney. It's the "Last twelve hours!" at Kohl's — "Hurry! Come in now for the biggest savings all year." The Sugar Plum Fairy celesta solo from *The Nutcracker* is the background music of every ethereal moment in commercials. Tiptoe through perfect houses, where every kiss begins with Kay: "'Twas the night before Christmas and Mom didn't know / That Dad was right behind her with a box, and a bow . . . This Christmas, tell her she's the greatest chapter of your story." Give her a necklace. Give her a cruise. Give her a Lexus. "Give your entire home that Febreze holiday freshness."

The TV drones on. Marissa stands above Caroll with a curling iron and a flattening iron. While Caroll drifts in and out of sleep, Marissa curls and then straightens her mother's hair, and then recurls it and flattens it again, over and over and over.

16

Wrapping

OVER IN the land of twinkling lights, the wait to get into Jeff and Bridgette Trykoski's cul-de-sac is longer than it has ever been. It's December 22. Families are trying to do it all, and time is running out. Someone tried to cut in line by turning onto Bryson Drive from the opposite direction, and "there was a New York–style honking war out here," Jeff reports. He had to go outside and get both drivers to knock it off.

A lot of the SUVs have tinted windows. All you can see is that the front and back seats are full. "Candy canes?" Jeff or I will offer, holding up one of the many plastic buckets of candy canes that Jeff gives away in a season, and the driver rolls the window down an inch, warily, as if you're about to solicit money. Some people always try to hand Jeff cash, holding tens and twenties out the car window. Jeff has never taken a penny.

Oh, this is your *house?* they ask. Yes, Jeff will reply, and remind them that there's a box by the mailbox for canned food donations.

Oh, we'll have to come back, most say, *and drop some off later.*

"Please do," Jeff will say, "and have a Merry Christmas."

Merry Christmas, they say. Some don't say much at all. A lot of people won't roll down the window, or they decline the candy canes with barely a nod. Others want to know everything: How he does it. Why he does it. How many bulbs is it, how long does it take to set up, does he hire day laborers to come to do it, what's his electricity bill for December? Jeff has so far collected 1,562 pounds of food, which he has stacked in the entry hall. It's an impressive amount, but he wants to exceed a ton. There's some elementary school that brings in the biggest amount of food each year to Frisco Family Services, and Jeff wants to beat that.

Bridgette's parents, Gary and Verna Iraggi, arrive, having driven several hours from their home in Victoria, Texas. Bridgette's brother and his family are coming, too. Unlike Jeff's folks, the Iraggis have accepted the fact that if you want to be with Bridgette and Jeff on Christmas, then you have to come to them. One of the first things Verna does, upon arriving, is adjourn to the guest room and shut the door, to wrap the presents Jeff has bought for Bridgette. "I outsource my wrapping every year to my mother-in-law," he says. Verna loves to wrap, and she likes to inspect what Jeff has bought, to try to figure out how much of it Bridgette will take back.

"Are we still going to Wal-Mart?" Bridgette asks Jeff, after ten o'clock. She's made a long grocery list. It's better to do it now than get caught having to go tomorrow, or on Christmas Eve. They get in the Honda Pilot and five minutes later are walking into the glowing behemoth of the twenty-four-hour super-sized Wal-Mart, the kind with a grocery store. Jeff gets a cart. "Is this list in any kind of order?" he asks her.

"Yes, Jeff!"

"Happiness, happiness," he says. "Let's have happiness."

It's beginning to look a lot like a paper-plate Christmas (plastic plates and bowls are on the list). They need French bread, three jars of spaghetti sauce, lasagna noodles, provolone cheese, mozzarella cheese, Parmesan cheese . . . One night they're having enchiladas and the next night will be lasagna. Christmas morning will involve an egg casserole. (Earlier, Bridgette and her mother endured the testy mob lined up at a HoneyBaked Ham in Plano, emerging with a spiral-cut glazed beaut.)

Wal-Mart is in a creatures-of-the-night mood, the fluorescent lights casting a cadaverous pallor on the faces therein. Stock boys run about cutting open cardboard boxes and stacking merchandise and bitching at one another in Spanish. Bridgette maneuvers the cart through aisles blocked with boxes, while Jeff heads off to make efficient sense of her list. He returns with frozen pearl onions, crunchy French-fried onions, cream of mushroom soup. French-cut frozen green beans, baby carrots, grape tomatoes, celery.

"There's no corn tortillas," Jeff reports.

"None?" Bridgette cross-examines. "None?"

Pepper jack cheese, some cans of enchilada sauce. A jar of minced garlic, a few bags of mixed salad greens, crescent rolls in the Pillsbury tube ("They're out, so get the generic kind, I guess," Bridgette says), and, back down the frozen aisle, it's the last box of microwavable Monkey Bread, which Bridgette tosses into the cart.

Ready-made Monkey Bread sends a sudden wave of loneliness through me.

There is, in fact, a Christmas I am missing, a holiday and a family to which I do belong. It's not with my own relatives, but it's what I have, and it is considerable. My boyfriend, Michael, and I go to his mother and stepfather's house every Christmas Eve. We stay the night, sleeping in what used to be Michael's bedroom. This would have been our fourth year of this routine, which would be more consistent, by far, than any Christmas tradition I've known since I was a kid. His mother's house is in suburban Maryland, a forty-minute drive from our apartment. On the way there, we always stop at Hunan Manor and pick up the Chinese food his mother has already called ahead for. We eat Christmas Eve dinner with his parents and, some years, a stepsister and her husband and kids. His mother has multiple Christmas trees (living room, TV room, dining room) and a small army of Santa figurines. It feels good there.

After his parents have gone to bed, it's Michael's and my job to make the "traditional" Monkey Bread for a big buffet brunch the next day, when several more relatives will be coming over. The reason we make the Monkey Bread is so Michael can feel like it's Christmas the way it always was. Or maybe this is his mother's idea. It's the same Monkey Bread she made when Michael and his younger brother, Matthew, were little boys.

Matthew was murdered a decade ago, on a college spring break trip in Florida.

I never met him.

Michael's mother keeps the recipe she clipped twenty-five years ago from an issue of *Woman's Day*, taped to an index card. It is faded and butter-stained. Even though the recipe is easy, we consult it every year:

You *schloompf* open a few tubes of Poppin' Fresh dough and roll it into little balls. Dip the balls in melted butter and then roll them in a gallon-size Ziploc bag filled with 2½ cups of sugar and 4 teaspoons of cinnamon. You fill a greased Bundt cake pan with little balls. You can add dried cranberries, or a package of Craisins. You put it in the oven, 45 minutes at 375 degrees.

It's usually close to midnight by the time Michael and I get our Monkey Bread in the oven. We watch an old movie while it bakes. One year it was *Meet Me in St. Louis*. Michael had never seen it. (Contrary to reports, not every gay man has seen every Judy Garland movie.) "Why is this a Christmas movie?" he asked me last year, a half-hour into Judy prancing around in that Technicolor spring.

"Wait for it," I whispered.

Soon enough, the holidays approach, the Smith family must leave St. Louis for New York, and glum Judy is singing "Have Yourself a Merry Little Christmas" to cheer up the bratty little sister.

Another year we just watched the broadcast of the Yule Log, mesmerized by a video of a fire, on the TV, next to the real but unlit fireplace. We got sleepy and curled up on the sofa, and as the video blaze crackled and split, the Monkey Bread burned, the smoke alarm went off, and we jumped up and ran around, waving oven mitts.

I was happy and did not know it until now.

It's not that there's no place like home. (Sorry, Judy.) I believe there's *every* place like home, even in a Wal-Mart in Texas. Here I stand, a few paces down the frozen-food aisle from Jeff and Bridgette and almost 1,400 miles from Maryland, thinking about the Christmas I left behind so that I could impinge on the Christmases of complete strangers.

"What else?" Jeff says.

"Beer," Bridgette says.

Through all this there has been the sound of another couple having an argument, a few aisles over. We pass a man in the aisle that divides groceries from women's apparel. He is about forty, pudgy, and he is sitting on a display model of a garden bench, crying. There's a wife next to him and four daughters, the oldest of whom is a very embarrassed-looking teenager in a black mini dress, torn fishnet hose, red Converse high-tops, and goth-girl eyeliner. "Well, I'm not going to stand here and be embarrassed," the teenager announces, stomping off, and the younger sister follows, but they go only about ten steps away.

"What?!" the woman asks her husband, who wipes tears from his eyes. "What is it? What's your fucking problem?"

"I just can't take this," he mumbles.

Bridgette is barely interested in rubbernecking this scene and pushes her cart down the potato chip aisle while Jeff goes and gets the beer. I am of course riveted. It's a full-on Christmas meltdown. It's a KLTY Christmas Wish gone David Mamet. I have to know more: Are they broke? Are they the Wal-Mart Cratchits? Is she going to leave him? "Are you eavesdropping?" Bridgette asks.

We come back down the next aisle and the wife is still harping on the man.

"Oh, my God," the teenager says, distancing herself farther from them.

"Frozen pizzas," Bridgette tells Jeff.

The man can't stop crying.

David Lynch could make a Christmas movie in this Wal-Mart tonight. An old woman's tan looks green in the hot fluorescence, and her makeup is smeared. People are missing limbs, or driving around in motorized scooters. We pass a man in the cosmetics aisle examining all the tubes of mascara. The Christmas carols are turned up full blast, almost torturously so: Boy George and Bono wailing, "Do They Know It's Christmas?" (Oh, do they ever.) "Let's get out of here," Bridgette says, finding Jeff and the full shopping cart.

"Happiness," Jeff reminds her. He whips out his credit card and pays: $137.81.

"I'm going to see Bryce," Jeff says, after unloading the groceries just before midnight. "He's waiting out in front of his house to catch kids." He drives four blocks into the next subdivision, which is called Prestmont.

Bryce Kindla's lights display is almost as popular as Jeff's. Bryce is also thirty-one, a computer tech guy, too, who develops software programs for online banking services. He and his wife just had their first baby. Each year Bryce installs a railroad track and an elf-sized train that choo-choos around his property line, with a large, lit Christmas tree in the center of his yard.

Bryce's house has been switched off for the night. Jeff parks the car on the street in front of the house next door. "Is he out here?" he wonders, getting out of the car.

A dark figure in a stocking cap, navy blue sweatpants, and a black fleece pullover emerges from behind the shrubs, carrying a baseball bat.

"Oh, hey," Bryce says.

"You're sleeping out here all night?" Jeff asks.

"Maybe," Bryce says. "It's not so bad when you get used to it."

The Christmas killers are out in force this year. Bryce has lost a couple of candy canes on his fence border. Jeff lost two of the mini trees that line the sidewalk in front of his lawn a few days ago, and his mechanical deer sometimes get rearranged into carnal poses. There is no way, Bryce has vowed, that he's going to let someone mess up his train tracks this close to Christmas.

People keep forwarding to one another a YouTube video of grainy night-vision security camera footage from the stabbing death of an air-blown Frosty the Snowman in a man's front yard, in some American suburb. The killer doesn't creep up on Frosty. He runs right up to him and stabs! (Stab, stab, stab!) Frosty never knew what was coming. The video led to an arrest.

Nothing gets Jeff and the other lights nerds more riled than vandalism. One friend of theirs set up a network of security cameras to catch a little boy peeing on his Santa doormat. It goes against everything they think they know about Christmas. Last year, Bryce caught a kid around the corner knocking down his "North Pole" fence. He went to the mother, "and she chewed him out pretty good and said he was going to have to pay for the damage, but his dad called the next day and said, 'Look, I had a real heart-to-heart with my son and he didn't do it.' And I'm like, yeah, right, I *caught* him doing it."

Bryce looks up — was that a car? He points to the houses where teenagers live. Some over there, "then those guys down there." A tricked-out Acura sedan with tinted windows drives by and things get skittish, rabbity. "This whole neighborhood is crawling with teenagers," Bryce says.

It baffles Jeff. "Why would someone do this to Christmas decorations? What could they possibly get from it?" he asks.

Right away, I can think of four things you could get from it. First, you get to be an asshole, with your asshole friends. (Never underestimate that.) Second, for some people, Christmas sucks. It just does. (So let's knock some stuff over.) Third, it's all so "cute" and "pretty." (So let's kill it.)

Last, there is perhaps some forgotten, dark-of-winter instinct that is so ancient and ingrained in Western culture that we're not even aware it's in us. The rowdy streets filled with peasants tormenting the rich homeowners. A man stabbing an inflatable snowman on YouTube is acting on our most primal winter solstice emotions.

It's a beautiful, cloudless, moonless night. Our breathing just barely makes wisps of vapor. "*Too* quiet," Bryce says, like Marines in movies always say right before Charlie's gunfire erupts from jungle cover.

'Twas two weeks before Christmas, our family was crazy
No time for each other, no time to laugh and be lazy

Dad was busy working with his Haggar sales team
His J. C. Penney account left him no time to dream

Mom had her Christmas decorating biz, 30 houses in all
She prayed off her clients' ladders, she wouldn't fall

So off to Vail, Colorado, the Parnells flew
For 5 days of skiing, tubing and fun they knew
— Tammie Parnell's 2006 Christmas card poem

Tammie calls. It's two days before Christmas and she has nearly all her shopping left to do. It's panic time. "Oh, my gosh, where are you?!" she shouts. "Did you miss me?"

"How was Vail?" I ask.

She tells me she got the "total moment" she so desperately sought, but not the way she planned. She thought maybe the total moment came one afternoon, at the top of a ski slope, as she, Tad, Blake, and Emily lay down in the snow and got their picture made. (That total-moment photo will be her Christmas card — which she'll get to finishing and mailing right away, along with the poem she's written.)

Several feet of snow fell on the day before the Parnells were supposed to fly home, in what became the Colorado blizzard of 2006. They sat in their rental SUV on the interstate from Vail to Denver as traffic slowly crawled through the mountain pass. The radio said all the flights were canceled. Tad took charge and got on the phone and called a Courtyard by Marriott. Through the blinding snow they found the exit. "And you know what? We were stranded there for two days, and I don't care," Tammie says, "because *that* was the total moment, because we were all together." Blake was worried they'd never make it home and Santa wouldn't be able to find him. (Despite what Cookie the Elf told him, he is still holding out every last shred of hope for a dirt bike.) The next day, the snow stopped, the sun came out, and across the drifts surrounding the Courtyard by Marriott, Tammie saw they were on the other side of the interstate from a Target and a row of chain restaurants, and she was pleased.

Tammie began networking in the hotel's lobby and complimentary-breakfast café among her fellow disaster refugees. A man who learned they're from Texas offered them four seats — for $2,000 — on the private plane he chartered to Tulsa. "Pay all that money and *still* have to drive home from Tulsa?" Tammie thought to herself. "Thanks but no thanks." Most of the travelers stranded there were Tammie's kind of people: affluent families who got stuck trying to get from ski resorts to the airport. After another day of dire reports from CNN and the Weather Channel, Tad made a final decision: keep the rental car and try to drive the whole way home, starting the next morning, the twenty-second, and just hope they make it. Now it felt like an adventure, something they were all in together, and of course they'd make it home. Of the trip itself, Blake said he liked the snowboarding best. Emily liked going shopping. Tad liked that there were no Christmas trees that needed to be decorated and that Tammie was calm. Tammie liked — nay, loved — stopping at a place where you can get pulled around on a dogsled by an ("absolutely phenomenal") team of huskies and begged everyone to come out and join her, but by then the blizzard was blowing too hard and too cold and nobody but Tammie wanted to do it. So out she went, in the blinding snow.

The Parnells are home now, and Santa Claus is on his way. After a furious day of last-minute deadline shopping on Christmas Eve, Tammie has made (ordered?) a dinner of ham, green beans, and scalloped potatoes, even though it's after eight before she gets it on the table. She is still operating under the delusion that she'll have time to bake homemade pies before it's officially Christmas.

Tammie, Tad, Emily, and Blake are sitting in their formal dining room, using the good china, and it is, briefly — in between Tammie telling Blake to please put his PSP away — a soft picture of bliss. Outside it's chilly and drizzly; inside Tammie has achieved traditional, domestic glow. The stockings are hung by the chimney with care. A plate of sugar cookies is ready for Santa on an end table by the living room fireplace, and Toby prowls and sniffs nearer and nearer to them. Tammie never got time to switch out the living room tree (decorated for her open-house Christmas merchandise sale weeks ago) for the more extravagant "family" tree she'd envisioned. She also never got time to go see Lorraine. There's a sign on the door of Lorraine's house telling visitors to just come on in, don't knock. "I just couldn't," Tammie says.

Now it is time for her favorite Christmas Eve activity — opening the cards. Tammie likes to set aside and save all the Christmas cards that come in the mail, insisting that they be opened and passed around by the whole family, all at once, gathered around the dinner table on the twenty-fourth. This is so Tammie, Tad, and the kids can "say a prayer for each and every one of our family and friends," she explains. "Isn't that right, Emily?"

Emily gets the basket full of cards off the kitchen buffet counter.

This is torture for Blake, who looks down at what his thumbs are up to in his lap.

"Blake, I am going to tell you one more time," Tammie says, "and then I am taking away the video games."

He sighs and stops, for only a moment. There are more than a hundred cards in the pile. As Tammie opens each, it seems she's more interested in dishing about each family than praying. For Tammie, talking about people *is* a form of prayer — I think she really does hurt for, and root for, each and every friend. These families have all sent pictures of themselves, as a group, or of their kids. The kids are all of course "phenomenal": phenomenal football players, phenomenal piano players, testing off the charts. Even the ones who have learning disabilities are phenomenal, *bless their hearts.*

The family that sent this card here is "really active" in the church's mission outreach programs in other countries, Tammie tells me. This one has a daughter on Emily's volleyball team. This dad survived brain tumors. This mom survived ovarian. This one moved to California to marry a new husband, who is a judge, and now she commutes back and forth because the ex-husband won't budge an inch on custody.

"Blake, take it into another room," Tad finally tells his furtively video-gaming son, and Blake is gone in a shot. Tad also wanders off, and Tammie and Emily spend another half-hour going through the Christmas cards — pictures of happy families in parks, or in front of fall-foliage backdrops, in matching sweaters, or, like themselves, on ski slopes.

"Okay, guys, it's getting late," Tammie finally says. It's after ten, and Santa's coming. They've checked online to see where the NORAD radars have tracked his sleigh so far. "Get upstairs," Tammie says. Emily and Blake want to sleep in the same room tonight. Emily brushes her teeth and gets in the other twin bed in her brother's room. Tad takes the tuck-in duty, coaxing them to relax, close their eyes, think of Christmas.

About twenty minutes later, he comes downstairs to find Tammie.

"Let's do the harp first?" he asks.

Tammie and Tad are in the garage, wrestling her aunt's restored harp up and out of the delivery crate. As they wrangle the harp into the laundry room, I turn and look back toward the kitchen. Emily has gotten out of bed, come downstairs, and is standing at the sink eating a banana, pretending not to spy on her parents.

"Emily?" Tad says, looking up from the crate. "What are you doing up?"

"I was hungry," she says.

"Get upstairs," he says.

She tiptoes away.

They decide to wait a while to bring the harp into the living room, where it will stand between the tree and the fireplace. Tammie has Christmas music on the stereo. "We have lots of wrapping to do," she tells me. "How about it, friend? You ready to wrap?"

Tammie, Tad, Toby the dog, and I go into the master bed-room and shut the door. It feels terribly intimate. In all this talk about Santa, I never in my life got to be him.

They have a debate about wrapping the presents every year, and Tad always loses. When he was a boy, the Santa Claus booty arrived under the tree unwrapped, because, Tad reasons, this is how Santa carries it around in his bag. In Tammie's girlhood, Santa wrapped everything in pretty paper — red for her, green for her sister Dee Dee.

Tammie wants to start with the stockings, which she goes and gets from the fireplace and brings back into the bedroom. "Jam stuff in all the way to the toe like this," she says. "Put some of the good stuff down deep. And fill it all in with candy." It's games and a DVD and lots of "pampering things" for Emily. It's games and a DVD and gross-out squeezable eyeballs for Blake.

"One, two, three, four, five, six, seven, eight," Tammie says, counting Emily's stocking stuffers.

"One, two, three, four, five, six, seven," Tammie counts in Blake's pile. "Seven. Do you think they'll count to see who got more? Will it make a difference, do you think?"

"No," Tad says.

Tammie disagrees. While I finish stuffing stockings, she and Tad disappear into the walk-in closet beyond their bathroom. They're back there a long time.

Tammie comes out with some clothes. "I totally jammed at Old Navy," she says. "This is 'To Emily from Mom and Dad'" — and that's my cue to grab a roll of paper, some scissors, and the Scotch tape. "Let's get busy, mister." She walks back to the closet.

"'To Blake, from Mom and Dad,'" Tammie comes back, announces.

I cut, I fold, I wrap. Toby plops down under the Christmas tree in Tammie and Tad's bedroom, next to me, and watching and then dozing. Every piece of Scotch tape winds up having a black dog hair or two on it, which I pick off one at a time. The radio is playing "We Three Kings of Orient Are," then Point of Grace, then Amy Grant, then "It Came Upon a Midnight Clear," then a soft country ballad version of "Angels We Have Heard on High" — "Gloooo, ooooooor, oooooria. In excelsis deo." Then, for what must be the millionth time, "Mary, Did You Know?"

I wrap a dozen presents, half for Emily and half for Blake, all tagged "From Mom and Dad."

At last we move on to Santa's presents, which involve the secret rolls of red and green paper, which Tammie is careful to hide away, separated from any other wrapping paper in the house or garage, as a cover for Santa. Tammie emerges from the closet again, with the Xbox for Blake, and the *Pirates of the Caribbean* sequel for his PSP. "Oh, you're caught up," she says, admiring the piling presents. "How fabulous is that?" She also got two Nintendo Dance Dance Revolution Extreme 2 games for Emily — "because you need two to play with another person."

She also brings out more clothes.

"Did Santa ever bring you clothes when you were Blake's age, or no?" Tammie asks, holding up some of the shirts that Santa is leaving Blake. Tammie kneels and starts wrapping some of the presents in Emily's pile. Red paper is for girls, Tammie says. Green is for boys. Tad comes out, looks at our work, and returns to the closet to sort through still more presents.

Tammie counts presents: only eight for Emily, but ten for Blake.

"So Emily needs two more, right?" I ask.

"Oh, we're just getting started," Tammie says, getting up, back to the closet, reappearing in a few minutes with still more — a book, a sweater, more games, more DVDs.

"You didn't know what you were in for, did you?" Tad says, stepping into the bedroom briefly again to look at the pile, then returning to the closet. Tammie keeps bringing the merchandise, and I keep measuring, cutting, folding, and wrapping, signing Santa's name in Sharpie on the gift tags.

Somewhere in this, Christmas occurs. I look at my watch and it's half past midnight.

Tammie's counting: "Fifteen, sixteen, seventeen, eighteen, nineteen — okay, that's nineteen Santa presents for Emily."

"Nineteen for Blake," I confirm.

"Santa is *done*," Tammie says.

"It's going to look like a scud missile hit here," Tad says.

We carry the gifts out to the living room, where Tammie and Tad arrange them into two piles on the floor in front of the sofa. "I told you," Tad says. "We go overboard."

Tammie's got one more gift — for me.

"Open it," she says, handing me a pretty bag stuffed with tissue.

She got me a green leather journal notebook, and deeper in, wrapped in tissue, is a small rectangular ceramic ornament on a red string. It has four words on it:

Believe in the Magic.

"Do you?" Tammie says. "Do you believe in it now?"

"I guess I have to," I say.

"You better," she says.

While I wrapped and wrapped those Parnell presents, the drizzle outside let up, and a beautiful fog rolled in. It's after one when I leave the gates of Stonebriar Country Club Estates, but I'm not ready to go home. I drive once more across the mall parking lots, completely empty now. The America that wakes up in a few hours will find itself celebrating one of the last, best Christmases in a long while, as far as one economy could take them before weakening into another reality.

Sitting at a red light I could easily run, I examine the trinket Tammie gave me: *Believe in the Magic.*

Easier said than done?

I didn't find the magic myself, but I found other people who claimed to find it everywhere they look. *That* is my Christmas present, pieces of a Nativity scene of my own assembly.

In Tammie I'd found a Mother Mary, who surrounds herself in the artifice of Christmas, sincerely moved by the beauty of plastic garlands, looking so hard for the total moment, and always about to miss it because she's so busy looking.

In Jeff I'd found my Wise Man, with his schematics and circuit boards and numbered extension cords. He is discovering new frontiers in holiday glitz, literally pushing an old Christmas to new limits, adding more and more lights to his house, and to the city of Frisco — and for what? Merely to make others happy, he told me over and over.

The real find here was Bridgette: tough and moody, yet brought to tears by unpacking certain tree ornaments. She has a healthy skepticism, bordering on scorn, for the princess/diva culture of the lives of women around her and triumphs over it by declaring herself the only princess who matters. To me she is an angel figurine, the one with the cracked wing that has been glued back together, wearing a sparkly top and Lucky jeans. Jeff and Bridgette never once spoke of religion; but what was going on outside, with the cars lined up around their cul-de-sac, was as close to a spiritual pilgrimage as I saw all December.

In Caroll, I had a found a quiet shepherd, tending to her three children, her flock on the hillside, which in our case had been a Best Buy parking lot. The shepherd is struck again and again by God's glory as the angels come down from the sky, heralding Pastor Keith's paths to enlightenment: "The real truth is always more important than our opinion of the truth," Pastor Keith said in one of his Leadershipology e-mails. And: "You can't always get what you want but you can always *be* what you want." (And: "The happiest moments in God's day are when you're happy." And: "You must embrace the benefits of dying before you can embrace the benefits of living.") Caroll took these little esteem vitamins and derived a nutrient from them I never could, even when I tried.

A little before 2 A.M., I stop and fill my gas tank at the 7-Eleven on Preston Road. The lady behind the counter says business has been dead for an hour, but they'll start coming soon. (December 25 is statistically 7-Eleven's busiest day of the year, when so many other stores are closed.) She worked last Christmas, and she knows: they'll start coming for doughnuts, for taquitos, for cases of Bud Light.

I walk back to my car. There's not a soul around, not another pair of headlights on Preston Road or in any of the parking lots. *All is calm, all is bright.* For a minute I bask in the peace on Earth and stare at the glowing logos: Applebee's, Target, Toys R Us, Red Lobster. I am certain I have it, whatever Christmas is. Then I think too much about it, and then it is gone.

American Greetings

(AN INTERLUDE)

A LITTLE BEFORE 6 A.M., Caroll's street is silent except for the chattering of grackles in the trees. She has decided to just leave the front door unlocked for me.

I turn the knob, stepping softly inside, following the glow from the Christmas tree into the living room. Santa brought Marissa a hot-pink mountain bike, which is parked and waiting for her in front of the TV. There are other presents he left in a neat row on the floor — tweenage girl things, jewelry things, pursey things, a Nintendo DS with the Nintendogs game, and a DVD of *The Devil Wears Prada*.

"Hi," Caroll whispers from behind me. She barely slept. She's in sweatpants and an old T-shirt, wearing her glasses and the boot brace on her right foot. She pours herself some coffee. "Marissa's pretty much awake, in my bed. I'm going to try to get Ryan up."

A few minutes later, he stumbles out, messy-haired and yawning, wearing red gym shorts and a gray T-shirt. He plops down on the floor in front of the love seat. "Now?" Marissa says, groggily, from the hallway, where she waits for her entrance.

"Not yet," Caroll says. She gets her camcorder ready. (Tape okay? Little red light?) "Okay," Caroll says.

"Now."

Now. I would like to point out all the careful notes I did take on Christmas Day, how serious it seemed, running from one house to the next. I watch as Marissa opens all of her presents and I write down each item she receives, which I will later cross-check against the home video Caroll made. I watch Ryan unwrap (and unwrap, and unwrap) the gift Marissa "got" for him and then mummified in several layers of paper and Scotch tape, which finally reveals itself to be a half-used bottle of macho body spray Marissa took from his bathroom counter. Caroll similarly receives an overwrapped present from Marissa, which turns out to be a ceramic candle Marissa purchased at the Celebration Covenant Church gift shop, with the saying "I am with you always," Caroll reads. (It's Jesus's parting line to his disciples in the Gospel of Matthew.)

"Yeah," Marissa says. "They had all these different ones, but I thought, I *am* always going to be with her . . . You said you didn't like candles, though."

"Nunh-unh," Caroll says. "I didn't say I didn't like candles. This is pretty." From Ryan, there are two presents to his mother. The bigger one is a universal remote control from Best Buy that will hopefully eliminate all the other remote controls in the house. The smaller present is a box of perfume. "I think he tried to get me what I asked for," Caroll says, peeling off the paper. "I don't get into all those expensive perfumes, but believe it or not the one I do like is at Lucky Jeans, but is this? . . . It looks different than the one I —"

"I went to the Lucky store but that's all they had," Ryan says. "I tried the Macy's, too. They changed it, they told me, but this is supposed to be the same."

When I leave, Caroll has a ham in the oven, and Michelle and Joey have arrived, and everyone's getting ready to watch *Rocky* on Blu-Ray.

Now. It can be reliably reported that Tammie spent most of Christmas Day flat out on the couch. It can be reliably reported that much of America spends the day this way, too. For all its buildup, Christmas itself is inert, lacking active verbs or anything like narrative arc, once everything is unwrapped.

Now. Bridgette unwraps her presents. She will return the Lucky jeans Jeff got her and the Victoria's Secret bra he got her. "I just love unwrapping bras in front of my family," she remarks, sitting by the tree, while, indeed, all eyes are on her. Jeff has helpfully — he thinks — prioritized the presents in the order that will "make sure your satisfaction is maximized," and, being Jeff, he narrates too much as she unwraps. "Needless to say, that was the least expensive gift, but the next gift is —"

"Jeff, I don't care!" Bridgette snaps. "Just let me open it in peace!"

It's the Coach purse, the one she'd asked for. Four hundred bucks.

But it's a different day now, and this isn't the purse she wants after all. It's going back to the Coach store for a different color and shape, something "that I'll use more," she says. Not everything is going back. She likes the Pandora charms for her Pandora bracelets.

Later, Jeff calls his parents, but his mother won't come to the phone. She's mad again they didn't come home for Christmas Day. "My dad said, 'We've got to figure this Christmas thing out once and for all,'" Jeff tells me.

* * *

We've got to figure this Christmas thing out once and for all. Much of who you wind up being, as measured in American culture, is determined in some way by what happened under your Christmas tree, this year and years ago, for good or bad, whether lavish or lowly. (If you never had a Christmas tree, then that fact shapes you all the more.)

It's torn paper, discarded bows. The oven set at 375. Some of us go off to church, to be reminded how off-kilter the priorities are. Later we unwind, drink some more, watch football, eat some more, reread the instructions to games and toys, call the customer help line in Bangalore. Some of us sulk, or try not to. The toy of the moment whirs to a stop, yet Christmas always works. Of all the notes I take today, nothing seems worth keeping on the subject of December 25, other than that.

Half Off

✦

(ONLY **364**

SHOPPING DAYS

TILL CHRISTMAS!)

17

Wal-Mart

I USED TO HAVE a pathetic, recurring dream about the Christmas tree in the living room of the house where I grew up, still blinking and decorated way past December 26. In the dream I am too embarrassed to let visitors inside, because then they will know the truth about us — the family that forgot to take down the Christmas tree. What could be the subconscious fear? An array of obvious symbols, in my case: insecurity about status, for one, or the convenient metaphor of a gay boy wishing to store the massive, glittery truth of himself away in a closet or attic. The dream might have been about my worry that we were not, as a family, holding it together. The year I turned fourteen, my father left my mother, and the tumult gave way to a sorry silence. That was the first Christmas we were able to get a real tree to replace our artificial Sears tree, because real trees made my father sneeze. My mother and I drove from Oklahoma to New Mexico that first Thanksgiving without him and sawed down a white fir, easily eight feet tall, at a managed preserve near a national forest. We tied it to the roof of our station wagon and triumphantly brought it home.

In the dream, Christmas is long past and I beg anyone who will listen to help me take the tree down. Sometimes I take it down myself, only to walk into another room and see that the tree has reappeared, fully adorned and covered with strings of tinsel. When I awoke, I would realize the tree had in fact been down for weeks — dragged out with our garbage cans and trucked off to a landfill. Yet I would have the dream through January and February and into the spring; once in a while I would have the dream in summer. Once in a far greater while, I wake up from having this dream even now, decades later. It can still take me a few seconds to shake off the abstract sadness and frustration. Christmas being over brings on the most heavy-hearted relief.

A posse from the Texas Christmas Lights Club musters online, in their members-only chat forum, during the holiday weekend. They are on a mission to find the most and cheapest leftover lights at the day-after sales. Each man has been scoping out his surrounding retail area, hitting every Home Depot, Wal-Mart, and Target to see what sort of inventory is left. Their plan is to hit the stores at dawn on December 26, to buy up as much as they can and then swap surplus lights with one another at their annual convention in May.

Jeff posted the original message six days ago asking for field reports, and since then he has been updating a spreadsheet of what each man found, and what everyone still needs. One guy hit a Wal-Mart in Farmer's Branch and a Target by Dallas Love Field airport; he reports 400 boxes of clear-100 minis, zero multi-100s, and zero in red and green, but 200 boxes of blue, and "plenty of icicles — clear, blue, multi — plenty of blue and clear [shrubbery] nets." Another man hit two Home Depots, two Lowe's Home Improvement Warehouses, and one Garden Ridge in east Denton County and reports zippo in overstock, nothing left at all. Another found a surplus of multi-100 minis at the Highland Village Wal-Mart in Dallas. A man in Corinth needs 60 boxes of 100-count minis in red and green and 24 boxes of clear and multi and can't pay anything over $1.15 a box, and "I'll tell you right now," he types, "I'm broke. Could seriously be March before I could pay someone if they get me everything above."

For his part, Jeff has been closely scoping out seven Home Depots, five Targets, and three Wal-Marts. So far his two most promising locations are the Target in Allen, east of Frisco, and a Wal-Mart in west Plano. This Wal-Mart winds up being his point of attack. This is the fancy new Wal-Mart, a store that opened several months earlier as an experimental prototype, beckoning the high-end customer who prefers organic products and hundred-dollar bottles of wine. (Plano and other suburbs of Collin County frequently act as testing sites for chain stores and restaurant concepts, which gives consumers here the occasional sense of being a chosen people.)

Jeff insists I need to be at his house no later than 4:50 A.M. if I am to come along, as the Wal-Mart opens at 6 and we are taking no chances. The year before, at a day-after-Christmas melee in a Target, a manager scolded Jeff for sprinting through the store toward the holiday aisles. He never slowed.

Christmas present is now another Christmas past, and here is Jeff in his SUV, headed for the west side of Plano a little after five. "We're only going for mini lights," Jeff says, which are mostly for friends who need them, even though Jeff will buy whatever he can find, since he will always have some use for an extra few thousand bulbs. At a minimum, he needs forty-seven boxes of green, twenty-five of red, and fifty of multi, which will now be marked half off. "I can always use multi-mini lights," he says. "It just really depends on the condition of the boxes." Jeff is convinced there will be a mob waiting, so it is with some disappointment that his headlights catch only a few cars in the Wal-

Mart parking lot when we arrive at 5:26 A.M., more than a half-hour before opening. "I guarantee you there'll be fifty or sixty people waiting in the next few minutes," he says.

Soon, a pair of headlights moves from one side of the Wal-Mart lot to the other — "See, they're trying to identify what door is going to open first," Jeff says. He would feel a whole lot more comfortable if we got out of the car now and spent the last ten minutes outside, ready to rush whichever door will open first. "At least it's cold. It feels like Christmas," he says. "Some years, the day after Christmas at Target, everyone's in shorts."

More people arrive, and at an agonizing two minutes past six, the doors open and about forty people — Jeff leading the pack — power-walk into the harsh light of aftermath, grabbing carts and heading in the same general direction of the Christmas junk. Jeff gets to the holiday lights first, and there are far more leftovers than he expected. We start loading up a flat pushcart with boxes and boxes of mini lights. Minutes later at the register, a clerk rings up Jeff's booty: $428.67 worth of mini lights, which, a day ago, would have been almost $900. It's about 40,000 bulbs, "almost as many as there are on my house right now," he says.

The city of Frisco held a contest a while back for schoolchildren (the only vocal environmentalists in town) to invent a nickname for a state-of-the-art recycling truck that travels the alleys behind their stout houses. Once a week, the truck picks up royal-blue, ninety-five-gallon containers with an automatic arm, empties them, and sets them back down, at a rate of about 1,500 homes a day. The winning contest entry, from a girl in fifth grade, was the name "Recyclina." The truck was then given a makeover as a sexy, eco-aware lady: Recyclina has big red lips on her grille and is batting her green eyes with mascara lashes. She wears a red bow atop her driver's cab.

It is two days after Christmas, a Wednesday morning, a little before 10 A.M., and thirty-four-year-old Derrick Brody is at the wheel of one of the Recyclina trucks, slowly making his way down an alley in the Plantation neighborhood. Derrick has to get out of the truck only every fifth house or so to pick up stray pieces of trash. Frisco generally produces about 900 tons of recycling waste each month. In November, the total was 924 tons. Boosted by the aftermath of Christmas, the total for December will increase to 938 tons. This does not count putrescibles and other garbage that will head another way, in another truck, to a Collin County landfill.

Recyclina churns and grumbles down the quiet alleys, alone in a sleepy village. Mechanical reindeer are stilled. Air-blown snowmen, Santas, elves, penguins, and Nativity creatures are deflated, their bodies sprawled on lawns *à la* the Jonestown massacre. Derrick, who lives on the south side of Dallas, has hauled trash for fifteen years. If Santa knows what you're getting, Derrick knows what you got, thanks to the stacks and stacks of empty cardboard boxes for new Dell computers, for Xboxes, for plastic kitchens for little girls, for massage chairs from Brookstone, for Fisher-Price's Tummy Time Together infant play sets.

Each time Recyclina is full (twice a day), the recyclable waste of Frisco makes an hourlong, twenty-eight-mile journey to the ass end of Dallas, to the Materials Recovery Facility (or MRF, known as "the Merf") operated by Community Waste Disposal (CWD), a private firm that holds several municipal trash contracts in the suburbs. Recyclina curves south along the LBJ loop and exits into a neighborhood of taco stands, cheap Chinese buffets, U-Haul lots, dollar stores, and strip clubs. Derrick drives Recyclina around the back of the Merf to the docking area, where she tilts back and excretes her morning collection into a thirty-foot-high pile, shot through with bright glimmers

267

of holiday red and green, faces of Rudolph and Frosty, and glints of snowflakes mixed in with familiar logos for almost anything you could eat, wear, play, or wash down a drain.

My walk through the Merf reacquaints me with the Spanish-speaking elves, now wearing blue surgical masks, standing over the streams of trash that pass before them on a series of conveyor belts, hand-sorting what the machinery misses and lets slip by. The contraptions are all painted in primary colors and pastels, which CWD managers tell me helps brighten the mood. (It reminds me of those old Warner Bros. cartoons where dog chases cat chases Tweety Bird into a cacophonous factory, with a Raymond Scott musical score pumping away.) The machines separate paper and cardboard from the stream, then plastic, then tin, steel, aluminum, and glass. Eventually the Merf makes large cubes of sorted trash, which are stacked by category: bales of plastic shopping bags, bales of plastic bottles separated by type and then smooshed together; bales of cardboard; bales of newsprint and other paper; bales of aluminum. I ask where most of it goes, once it gets trucked out.

"China, a lot of it," says Robert Medigovich, the municipal coordinator for CWD, who also lives in Frisco.

China, I repeat. What's salvageable of this Christmas goes back to whence it came, to the real North Pole, where even now, the stuff of next Christmas is being manufactured and packaged up for a trip back to us, next Christmas.

The rest is garbage. Frisco's nonrecyclable refuse goes in a different direction, twenty-two miles east on State Highway 121 to the North Texas Municipal Water District's regional landfill. On the morning of December 28, a friendly water district engineer and spokesman, Jeffrey Mayfield, drives me out there. Along the way, he gives me a brief primer on the history of the landfills north of Dallas. He also talks about the long-term scenarios for the area's waste and the rate at which the garbage increases each year. He speaks of such things as *geomembranes* and *bentonite levels*. The old landfill in McKinney was now 99 percent full; the new landfill, farther east, opened in 2004 with a capacity of 70 million tons of garbage — enough, it is hoped, to clean up after a few million North Texans for another hundred or more Christmases.

Mayfield's truck climbs a mile or so on a switchback road into the desolate realm of the new landfill until we reach the active dump, where hundreds of gulls swarm overhead. The Merf had offered me brightly colored machinery and busy activity, with the symmetrical hope that what goes around comes around. But the dump is a dump. We don hardhats and tromp about on its mushy gray earth, where pieces of trash poke through. A stream of *blechh* falls out of an upturned dump truck, and in this, already, is a Christmas fir, stripped of its ornaments and ribbons and lights. It tumbles down a mountain of everything we did not want.

For every binge, a purge. The American consumer is now busy returning and exchanging unwanted gifts, cashing in gift cards, and pawing through reduced merchandise, tapping into the same sleep-deprived frenzy that brought out the dawn doorbusters on Black Friday not even five weeks ago. Only now it has the patina of unrealized hopes (in retailers) and a ruthlessness (in shoppers *and* retailers) that is no longer contractually bound to be part of some sweet story we tell ourselves about the reason for the season. With Christmas over, the "Xmas" pathogen can at last show itself in the autopsy. But is the patient ever really dead? The general complaint, even among people who love Christmas and every last thing about it, is that it starts too soon. Economists say the real trick now is not to pinpoint the *beginning* of Christmas, but the *end*. There isn't an end, at least not on paper, not in overseas factories, not in retail marketing departments, not in the film and music industries, and not in the shipping news.

What Americans spent on Christmas cannot be accurately measured until February or later, when enough of the ever more ubiquitous gift cards (nearly $30 billion worth this season) are redeemed. In ledger-speak, gift cards are nothing more than a liability against profits on future sales; if they expire before they are redeemed, the consumer loses, but in a way, so does the retailer, left with surplus inventory. TowerGroup, a retail research firm, estimated that a total of $8 billion in gift cards in 2006 were floating around unspent.

Whether or not gift cards make us truly happy on Christmas morning is still up for debate. In some families, the gift card reduces the entire experience of giving one another gifts under the tree to the handing over of plastic cards, many of them for the exact same amount for the exact same stores. Gift cards confirm the very sameness that New Urbanists and some city dwellers distrust about the suburban existence, acknowledging that we really do all have the same stores and restaurants, no matter where we are, living without variety or surprise.

"The whole character of the season has changed," says Michael P. Niemira, the chief economist of the International Council of Shopping Centers, who has been watching Christmas sales since the early 1980s and is one of the most widely quoted economists during the holiday season. It seems to him now as though Christmas never quite ends. Niemira tells me he is surprised in 2006 to see reports of people who get on their computers Christmas morning, redeeming gift cards at retail websites the minute they receive them: "It seems like, gee, isn't there a day anymore that we don't shop?" On January 3, Niemira's numbers shake out to a 2.8 percent gain from last year's Christmas sales, which, although an increase, is "more pessimistic than optimistic," he tells the Associated Press.

The minute a Christmas season subsides in the United States, it begins again in Chinese toy and decoration factories. Containers of ornaments and decorations bound for America begin leaving China on ocean freighters in the late spring, and the shipments peak in the earliest part of fall. By then, decorations are up in American stores and the carols start playing in October. Christmas TV commercials begin as early as six weeks before Black Friday, as does the arrival of Christmas catalogs and movie trailers for holiday-themed comedies. Pop stars book studio time in May and June to remake Christmas standards they hope will sell for seasons to come.

Jeff and Bridgette spent about $2,000 on Christmas gifts for one another and their families and friends in 2006. They spent another few hundred on parties and dining out during the holidays. Jeff is willing to tell me everything about the Christmas lights on his house except how much he spends overall on the spectacle. "I just won't discuss it. This is my only hobby and I don't ever talk about how much I spend on it," he says. When the Trykoskis' Visa bill comes in January 2007, Jeff pays it in full. (Like any good numbers whiz, he refuses to get caught by compounding interest.) Since Jeff began his own lights display consulting business, the coming and going of Christmas lights is now a year-round affair for the Trykoskis, involving thousands of dollars in inventory. He is also now licensed by Light-O-Rama to sell its products. Pallets of Christmas lights and other supplies are always arriving at the house, in transit to an increasing number of customers.

For being so revered for his holiday lights and spirit, Jeff exhibits no sentimentality for prolonging the season. On his website, www.trykoskichristmas.com, he has advertised that he will keep the lights display at his house on every night up until ten o'clock sharp on Friday, December 29. Bryce Kindla — the guy with the Christmas choo-choo train and the baseball bat to guard it — turned off his house a couple of hours before Christmas was even officially over on the night of the twenty-fifth. By midnight, Bryce gloats, pieces of the train were already boxed up and in the garage. In another couple of days, he and his wife and baby son are off on a trip to Aruba.

Jeff admires this. Even December 29 feels too long to wait. For all their excitement about the arrival of the holiday season, the members of the lights club truly geek out on putting it all away, neat and orderly. They are not beyond showing one another pictures, online, of their tidy garages and attics. "Oh, ten minutes after I switch it off, I'll start taking stuff down," Jeff says. And lo, on Thursday, December 28, Jeff shuts down his lights show a night early because of a rain forecast (putting away wet lights is out of the question), but also just because. Indoors, Bridgette is equally efficient with the end of Christmas. The snow villages go back in their Department 56 boxes and are stored on the high shelves in the master bedroom closet. The ornaments come off the tree, and the tree is put out in the alley with the trash. The snowman shower curtain is put away and replaced with the normal shower curtain. "I want it gone," she says.

On January 2, Jeff, Bridgette, Greg, and I meet at Frisco Square to unplug the big show on which Jeff and his brother had worked so many demanding hours to perfect. The management at Frisco Square enthusiastically counted some 80,000 to 90,000 cars that drove through the property between Thanksgiving and New Year's and has already signed Jeff to a contract for an even bigger show next Christmas. While we stand on the square and wait for Greg to drive up and meet us, one woman sees Jeff and gets out of her car to ask him if he's "the guy" who does all this. She tells him the lights show had made her family's entire holiday. Jeff thanks her, but what he doesn't tell her is that the lights had made his.

"Yeah, and we're going to Jamaica in March," Bridgette tells me. "Jeff has to take me on a vacation now."

Greg walks toward us.

"I like how you're wearing flip-flops and it's January," Bridgette tells him.

"When you have forty pounds of extra blubber it doesn't matter," Greg says.

We ride up the elevator to the fifth floor, to the conference room where Jeff's computers are still running the show on autopilot. "All it takes is to push one button."

"I want to push it," Bridgette says.

"Are you going to say anything profound?" Greg asks.

"No," Jeff says.

He checks his watch and pushes the button himself, and like that, Christmas is off.

Christmas has also been swept clean from the stage at Celebration Covenant Church, making way for a stark metallic gray set for Pastor Keith's sermon series "Church Royale 2007," taking as its inspiration the recent James Bond remake, *Casino Royale*. In this seven-week series, Craft's sermons promise to "lead you on an adventure that will impact every area of your life . . . to shift your life into 'Xcel-aration.'" Pastor Keith's sermons in this series will get 007-style themes: "In His Secret Servant-Leadership," and "The Living Daylight Saving Time," and "The Spouse Who Loved Me."

Caroll and Marissa Cavazos and I go to Chili's after church. "It's almost like [Christmas] didn't happen," Caroll says. She spent about $1,400 on Christmas presents for her kids, some friends, and a few relatives, in addition to another few hundred on Christmas food and the trip to Oklahoma and back to see her mother. She put all of it on her Bank of America card, and when the bill comes in January, she pays it in full from her holiday savings account. For a New Year's resolution, Marissa has chopped off almost a foot of her long hair for Locks of Love, a charity that claims to weave donors' hair into wigs for disadvantaged children who have cancer or other diseases that lead to hair loss. (Most of the hair collected by Locks of Love is thrown away, or sold to pay overhead costs, reports the *New York Times*. Does every pious bubble have to burst? Must I sit on this secret, too, along with my knowledge that some of the angels on the Angel Tree aren't real?)

When we get back to the house, Marissa begs me to come outside with her and watch her ride the new pink bike Santa Claus brought her on Christmas.

"She won't go out and ride it by herself," Caroll says. There are very few children left in the neighborhood, not at all as it was when Caroll's other daughter, Michelle, was little. "I keep telling Marissa it's okay. She went out and rode it up and down the street and came right back, and I said, 'See? You can do it.' But she just won't go out there, because of" — and here Caroll quietly tilts her head east, toward the top of the block, so Marissa can't hear.

"Because there's a child molester," Marissa says loudly.

Caroll had gone to a website that tells you if there are convicted sex offenders in your neighborhood. Sure enough, there's a guy at the top of their street. But the site doesn't give you details, circumstances. This is peace of mind?

Marissa and I go outside. While she pedals up the street to the cul-de-sac, I jog alongside her. She points out the house where he lives. "Have you ever seen him?" I ask.

"No," she says.

We stop at the top of the street, where a dirt path leads to a long swath of grassy field along which there is a row of mammoth utility towers strung with power lines. The ten-year-old in me would be all too willing to pedal on, into the field, as far as I could go, see what's out there. Marissa turns around, back toward the house.

When we walk in, Ryan is making his lunch in the microwave. Caroll drags a cardboard box and a plastic tub filled with unsorted family photographs from the entry hall closet to the living room, because I'd been asking about past Christmases. Marissa starts dumping Wal-Mart photo envelopes on the floor. Ryan gets interested and starts looking with us.

"Is this me?" Marissa asks, holding up a picture of a baby in arms.

"That's your sister," Caroll says. "Here's you."

"Baby Marissa!" she coos.

Here's an old photo album that Caroll's mother sent a while back. Here's Caroll as a teenager in Oklahoma City. "Me in one of my ugly states," she says. Here she is at seventeen, right before she got married to her first husband.

Here's Okinawa in the 1950s, when her father was stationed in Japan. "This is your Uncle Dodo," Caroll says to Marissa, pointing out a picture of a little boy in a black-and-white snapshot, but Marissa and Ryan are preoccupied with a more recent stack of color pictures. Caroll becomes momentarily absorbed. She stops turning the fragile pages of the old photo album. "This was the house," she says quietly. The house in Topeka . . .

"Here's the Christmas that Granny was here," Ryan says, at last, holding up a picture from six years ago. Marissa snatches the picture from Ryan: "It was when I got my roller skates, the blue ones."

"And I got my drum set," Ryan says. "That was, like, ninth or tenth grade —"

"My Fisher-Price medical kit!" Marissa suddenly screams, seeing herself at age five with her new toys. "Mommy, what happened to this? Do you remember?"

"No, it was ninth grade," Ryan says. "Here's —"

"Mommy, I can't get this open," says Marissa suddenly, as if sensing that Caroll, who is tracing a finger over one of the old black-and-white shots, is going to say something about what she's seeing in the pictures, something secret and grown-up, something gone and yet still around. Like that, the mood changes. Caroll is not listening to Marissa or Ryan.

It's the house they had in Topeka. "It's weird because this is the first time I've thought about this house —"

"Or it was tenth grade," Ryan says, flipping through pictures fast now.

"Because this is the house where, I was, where it . . ." Caroll pauses. Now we are talking about a different kind of shadow on the wall, not the house in Okinawa, not the rocking horse memory. This was Topeka. *This was the house.*

Looking at the picture, Caroll tells me about the memories she had buried, about her father's abuse. "I didn't know it until I was in my mid-thirties. It was repressed." The pictures get to the heart of it, she tells me. The heart of what? Caroll's fears, and her worries that she's not good enough, the "high responder/low doer" personality-test results, and those dark, occasional feelings that God blesses others while passing her by. That's the shadow on the wall, really. That's what she had to face: doubt, disbelief, unworthiness. She never confronted her father. When he died, she stayed out of the fuss over the will. She hoped if there was anything left coming to her, it would pay for her "first fruits" tithe at church with enough left over for Lasik surgery — her search for better, clearer seeing. When the money came, Caroll says, it paid for the operation almost to the penny.

"Mommy, you're having an emotional moment," Marissa says.

"I am," Caroll says.

Marissa curls into Caroll's side, now interested in the black-and-white photos, which, as Caroll turns another page or two, become the fuzzy Kodak colors of the 1970s, a teenage girl in a bikini. "Oooh, a swimming pool! I want a swimming pool! Mommy, you're cute there!" Marissa says. "You had blond hair?"

It's getting further away now, with each page that Caroll turns.

"Mommy, I don't like it when you're wearing those glasses."

"It was either ninth or tenth grade," Ryan says, still mulling over his Christmas drum set.

"He still has them," Caroll says of the drums.

"This was Hawaii!" Marissa says.

"No, Marissa, that's a pool at an apartment complex. Where your sister used to live."

"Do you still play the drums?" I ask Ryan, looking up at him from the photos.

Ryan seems about to answer my question about the drums.

"Not anymore he doesn't," Caroll says, answering for him.

The search for Christmas pictures ends at the bottom of the tub. Caroll thinks we got nowhere. "I know I took pictures; every year, I took pictures," she says, a little frustrated. "It seems like they should be in here. There should be a lot more Christmas in here."

18

The Epiphany Party

KELLI CALLS TAMMIE with the news five days after Christmas: their friend Lorraine has died. Tammie and Kelli have a confession for one another: neither of them went to see her, not since the night they'd gone over and put up her Christmas decorations in mid-November.

Kelli had been picking up Lorraine's son and daughter most mornings and taking them to school and dropping them off in the afternoon; Lorraine's father-in-law waited at the door. But she never went inside. She cried when she told Tammie that she just couldn't take seeing Lorraine like that anymore. She wrote her a letter instead and dropped it off at the house on one of those afternoons. Tammie understood. "I said to [Kelli], 'You know someone read that letter to [Lorraine], you just know,'" Tammie recalls. "'You took her kids to school every day. You've got to let that go. We totally decorated her house! We brought her so much joy.'"

The funeral is at a big evangelical church in Frisco on January 3. Several hundred people, including Tammie, are there. For all the cancer stories that go around (I've lost track of how many times the subject of cancer comes up in the dozens of interviews I've done around Frisco since September), Tammie says this sort of death had not yet happened in her immediate circle of friends. It has a chilling effect. She tells me she can feel another one of her epiphanies coming on.

Taking Christmas down is not a service Tammie offers her clients. She had given me a funny look weeks ago, when I asked if her rate included coming over to the client's home and putting everything back in boxes, wrapping ornaments back in tissue paper, and so on. "Gosh, no," she'd said. "Can you imagine what that would even be like? I'd lose all that time and money." (But this did lead to an interesting complaint: when Tammie's clients have their cleaning people or nannies put everything back in the Rubbermaid tubs, "you can totally tell next year, when I have to get it out and see that it was put away real sloppy, and you just get the feeling that whoever did it really did not care.")

One of the first people to get her own Christmas stuff up in the fall, Tammie is one of the last to put it away in January. She's an Epiphany girl all the way, keeping her living room tree and garland up and rearranging her Nativity scenes, waiting for the feast day of the visit from the Wise Men.

The same way Jeff has been buying up leftover Christmas lights, Tammie has also been a busy scavenger elf, hitting the hobby and crafts stores: "I drove to three or four Michaels and got some *phenomenal* stuff . . . I think I spent $200 or $300 and I have about seven big tubs full. I won't have to do anything next year. You're paying, like, 10 percent of the original cost. I got all these great wreaths and berries and great, great stuff. My car is loaded down, baby." At Tammie's favorite Hobby Lobby, the one in Plano, "this woman got ahead of me there. She must've had ten trees on a cart. And I said, 'Do you really need them all?'" By New Year's, Tammie has acquired seven new trees, all of which are destined for her attic, to wait out spring and summer. "Tad's saved tax returns from fifteen years — outta there! I want my attic to be clean and I need the room. I mean, *seven* trees. I bought one seven-and-a-half-footer, for like twenty bucks, prelit, multi-colored lights. I am telling you. I was *on* it."

The Parnells spent thousands of dollars on Christmas in 2006; Tammie is unsure of the exact amount. She also earned thousands in holiday-related cash, grossing about $27,000 from twenty-nine paying clients. She guesses she spent about $7,000 on supplies and décor, reselling her finds to clients at a slight markup, and she hasn't yet figured how it all shook out. The five-day family ski trip to Vail cost a little more than $5,000, including the extra fee for driving the rental car all the way back to Texas. Tammie and Tad spent about $1,200 on gifts for Blake and Emily, and another few hundred on gifts for one another. Tad gave her the *Alias* boxed set and . . . what else?

On the night Tammie and I meet at a wine bar to enumerate and go over the details, she can't remember what else she got. (And don't ask her what she got Tad; for years that's been a sore subject. "I can never buy him anything he wants," she says, and she's all but given up trying.) The Parnells spent another $500 or so on gifts for relatives and friends. They spent $200 entertaining, which was basically the Hotties party, unless you're counting their December country club tab; then it's a little bit more. It's difficult (if not entirely pointless, Tammie says) to sort out her family's holiday spending from the rest of the year. "It's always Christmas around here," she says. For his birthday in February, a few weeks from now, Blake will wind up getting that dirt bike he wanted. Emily will continue to wish for her own laptop for another two years.

The feast of the Epiphany, January 6, is on a Saturday this year, and Tammie gets excited about hosting what has become her now-annual afternoon Epiphany party for about twenty-five women. She serves punch, Champagne, finger sandwiches, and desserts. Kelli and some of the other Hotties are invited, but it's mostly women Tammie knows from her church, St. Andrew United Methodist, or from Blake and Emily's school. Tammie instructs her guests to wear white and gold and to come with a single, personal word in mind — an "Epiphany word," as Tammie describes it—which will express some idea or theme they want to focus on in 2007.

Sitting in Tammie's family room in the afternoon light, the women settle into the sofas and chairs looking like an ethereal gathering of bottle-blond and gently coiffed angels. They talk about Lorraine's funeral, which many of them attended, and Lorraine's kids. The idea of leaving their own husbands and children behind — and what awaits any of us beyond death — is at once an unthinkable and irresistible narrative to imagine. Tammie gets up and pushes Play on her stereo, where a CD is cued up. We sit and listen to Point of Grace sing, over and over, "We're not that far from Bethlehem."

Tammie's friends go around in a circle and share their focus words:

Strength is one.

Another woman says *calm*.

One woman shows her pink rubber bracelet (breast cancer awareness) and says she's wearing it on her right wrist and that her word is *look right*. "*Look right,* because then the bad things are *left*." (I want to slap my forehead but don't.)

Tammie is ready to tell us her word.

"My word is *listen*," she says.

The women laugh.

"I know!" Tammie squeals. But Tammie insists that listening has been her real epiphany. She is fully aware that she interrupts everyone before they finish a sentence. She knows she's a show dog. She doesn't want to be like that all the time. She reads from Psalm 46: "Be still and know that I am God." Tammie says she wants to calm down and hear what people around her are saying — Tad, her children, her friends. "I am saying to y'all, to remind me that this is my year to stop and listen."

Eventually it's my turn. Tammie insists I share the word I came up with for my year ahead. I tell the women my word is *finish*.

"Completion! Closure!" Tammie interjects, deciding on better words for me, not two minutes after she told the room it is her intention to *listen*.

"*Finish*," I say.

Five days later, on Thursday, Tammie and Kelli let themselves in with a key to the front door of Lorraine's house. Nobody's home — Lorraine's husband is at work, and the kids have gone back to school. It's dark and quiet. They've come to take her Christmas decorations down, the same ones they put up almost eight weeks ago, while she'd watched. In a little less than two hours, they pack up all the ornaments, disassemble the tree, wrap up the garlands, and put it all away in boxes to be stored in the garage.

One of the last things Lorraine told Tammie was not to bother with the garland over the fireplace in the family room. Lorraine had wanted to do that piece herself, with the kids' stockings. This morning, now, in the family room, Tammie notices that garland never did get put up.

Finish. The January skies darken and the weather turns colder than it ever was in December. An ice storm blows hard across the prairie. On a night in early January, I unlock the front door to the house at the top of the cul-de-sac, where I've been renting the downstairs bedroom for the last five months from the former Marine and his cute girlfriend. I'm greeted by an odd sight: their artificial Christmas tree is half taken down — the top part of its plastic trunk is on the dining room floor, and plastic branches have been plucked from it and are scattered in the foyer, as if someone had abandoned the task midway through. The next night, I find her packing several boxes in the living room. He's upstairs with the door to the master bedroom closed. She whispers to me that it's over between them and she's moving out. A day later, she takes her washer and dryer, her sofa, her two cats, and the rest of her things.

A couple of mornings after that, in the kitchen, I close the refrigerator door and am startled to meet a new housemate standing there, a man in a tank top and shorts, who's rented one of the other upstairs bedrooms, arriving here the same way I arrived, as a stranger via Craigslist. He shakes my hand and tells me he's a manager at a Planet Tan. His skin gives off an orange, Planet Tan hue.

The stricter homeowner associations have sent out letters, gently reminding residents to take down their Christmas lights and wreaths, the lamppost and mailbox ribbons, the manger scene figures, the Charlie Brown and Mickey Mouse lawn cut-outs, the air-blown Frosty Snowmen, and all the rest of it no later than January 15, if they haven't already.

The letters get less and less gentle now. On the second Friday in February, I pull my car over on a street in one of the subdivisions, put on my blinkers, and step out into the wind to take a picture of three lonely, weather-beaten plastic Christmas wreaths hanging on the iron fence that surrounds a fake lake. A Frisco police SUV rolls up as soon as the second Polaroid slides out of my camera. "Can I ask you what you're doing?" the officer says. He tells me I'll have to move my car, and that I shouldn't be taking pictures anyhow, without permission.

Finish. I drive off the next morning, up to the very edge of this world, where the Dallas North Tollway stops at a traffic light and becomes a two-lane road.

And I go away. I make a left on U.S. 380 and drive west, far-

ther and farther, across the country, seeing Christmas in reverse.

I drive from Texas to California, where, from a distance, I
watch super-sized freighter ships from Asia dock at Long Beach.
Cranes lift off the railroad-car containers that are filled with
who knows what, coming to a mall near you.

In early March, I take a week or two to drive all the way back
to the East Coast. I stay only in the newest, most prefab Execu-
Comfort-Suite-Homewood-Stay hotels I can find, the ones
nearest the box stores, chain restaurants, and shopping centers.
On the interstates, I exit every time I see a year-round Christmas boutique, emporium, or shoppe, taking a moment to walk
among refrigerated forests of blinking display trees. I stop and
stare into the plastic eyeballs of mystical creatures.

Baby, Please Come Home

19

The Container Store

THE ELEMENTAL STORY never changes: A man and a woman are worried about a baby. They return to the suburb of the man's childhood. They are on the road, and exit at Bethlehem Heights, and find the Comfort Inns and the Courtyard by Marriotts all booked. It is a story of crisis. It is set in Judea, in Whoville, in Bedford Falls, in the newly zoned nullity.

Ten months after Christmas, the nine new miles of the Dallas North Tollway open along Frisco. The bones in those two forgotten pioneer-era coffins, dug up from the slate-gray earth at what is now the Lebanon Road interchange, are never identified because there isn't enough DNA material left to test. An archaeological consulting group submits a lengthy report to the North Texas Tollway Authority in the summer of 2007, detailing what was found: "Burial 1 consisted of the coffined remains of an infant," according to the lab analysis. There was a scattering of bone fragments in the soil, barely enough to hold in cupped hands, including, the report reads, "portions of the squamosal, lateral and basilar portions of the occipital; the petrous portions of both temporals; several small fragments of cranial bone; neural arches of the first and second cervical vertebrae . . . 17 miscellaneous neural arch fragments . . . 11 unsided rib fragments . . . The developing crowns of four deciduous teeth

. . . consisted of the mandibular central incisors, the maxillary first molar and the left mandibular first molar . . ."

The bones are categorized and labeled in separate plastic bags, which are stored in a container on a shelf in a lab at Baylor University in Waco.

That story ends there.

Of course I go back to Texas and find Frisco always pregnant, expectant, abloom with signs that say *Coming Soon.* A Super Target is almost finished at an exit off the new tollway stretch in a power center that has yet another Best Buy and yet another Chili's and space for much more. "New *location.* New *relationships.* New *opportunities"* reads a sign in front of yet another new Bank of America branch. At least three mixed-use projects are depicted in the PowerPoints and presentation boards of unflaggingly optimistic developers. One of these is called Newman Village, which will have 900 "high-end" homes, 270 townhomes, and 900 condo units surrounding a pond, an outdoor shopping mall, and an amphitheater. The Newman Village land had been farmed by John Lennox Newman, one of the original settlers in 1841, and by every generation since, until it was the last working family-owned farm in Frisco. Newman's great-great-grandson, Jim Newman, sold the land to develop it for residential and retail use. "I came from the dirt," he told *Frisco Style* magazine. "I played in the dirt. I farmed the dirt. I sold the dirt. I developed the dirt. Someday I will return to the dirt. I guess what I know is dirt."

Where the Home Depot is now.

Where there were just cows.

I go back for part of one Christmas season, in 2007, and then for another in 2008. I go back at other times, for almost any reason, sometimes just to graze with my Capital One card at

Stonebriar Centre and shop so much that I have to buy cheap duffel bags to bring home my mindless booty of Gap cargo shorts, DSW loafers, Kiehl's facial balms, J. Crew "summer-weight" cotton chinos, and American Eagle Outfitters hoodies, in two colors, all of it marked down.

I go back and accompany Caroll and Marissa Cavazos to Celebration Covenant Church, where Pastor Keith tells us one Sunday that "we live in the most prosperous time in American history, the most prosperous, where you can pay $3 a gallon for gas — not that you like it, but the point is you *can*. Let's give a big hand for that — amen! — and you say, 'I'm not the most prosperous I've ever been.' Well, stick around. You shall be." He is relentless and Caroll needs him to be. *Think* noble! *Be* grateful! *Do* generosity!

On a different Sunday, Pastor Keith takes note of another new sign from above: three of Frisco's Starbucks are closing. (Starbucks is closing 600 U.S. locations, and this event is portrayed everywhere as a harbinger, a caution to our pampered selves.) "This is because people aren't spending $5 on lattes, they're going to be spending $5 on gas," he says. "But we are not going to let that stop us." Work on the $24 million Celebration Covenant cathedral complex will break ground soon, Pastor Keith says. There has been delay after delay. The latest snag has to do with the soil survey on the empty land behind us, something about stability issues twenty-six inches down.

On the freshly paved (or about to be paved) surface of things, everything seems the same at first. A new Hobby Lobby opens, with lighted holiday garlands and extravagant angels displayed in late summer. Then comes a new Container Store. Both are greeted with delight.

Then it slows down. Something stops. You don't have to be

an economist to read the data and feel some vibes, to discover stability issues twenty-six inches down.

The foundation on which the new century's economy has been precariously overbuilt begins to lean and then wobble in 2007. The Visa-wielding American shopper, so used to being thought of as a world champion, swoons in 2008 and begins a series of collapses heading into 2009. In almost every measure of the economy, the Christmas of 2006 that I'd spent in Frisco increasingly has taken on the nostalgic quality of the last "big one," when people spent money as they pleased, before the contractions started, before the foreclosures, before the price of a gallon of gas doubles and then drops unsettlingly to a price lower than it had been all decade. American consumers had borrowed $350 billion in home equity loans or second mortgages in 2006 alone, acting as though money would always be there to burn. A year later, creditors start sending up an alarming series of flares. It turns out too many of us have been sliding *imaginary* money around, believing it to be real. Stocks fall by a third. Job losses stack up in the hundreds of thousands each month.

Housing bubble, equity bubble, credit bubble, SUV bubble, dollar bubble, China bubble, Euro bubble, airline bubble, retail bubble, manufacturing bubble.

Coming Soon.

Some other things happen, too. The story keeps returning to where it begins: a man and a woman on a journey are worried sick about a baby.

Caroll Cavazos's grandson is born on April 24, 2007, almost two months early. He weighs more than six pounds, but most of that is fluid being retained by his terribly swollen heart. Caroll's daughter and son-in-law name the baby Van Joseph.

Van is transferred to Children's Medical Center in Dallas, where an array of life-threatening issues is immediately diagnosed. Michelle and Joey are thrust into a world of crucial decisions and staggering costs, which their insurance will cover up to, but not beyond, a couple million dollars.

After many days at the hospital, Joey starts to gets antsy when other family members are around. Caroll understands this completely and chooses to take longer lunch hours on Tuesdays and Thursdays to drive down to Children's, when her daughter and Joey aren't there. She can watch Van but no one is allowed to touch or hold him. An acquaintance from church calls to check on her, and Caroll gives her what she thinks is basic information, without candy-coating the odds: Van may very well die. "She said, 'You can't talk like that, you can't think like that,'" Caroll recalls, "when all I was doing was just saying what was really happening, what the doctors were saying to us." The woman insists Caroll banish "all the negativity" and just believe, just stay positive, and for once, this is something Caroll does not want to hear.

Caroll and Marissa go to the hospital a lot all summer, where Marissa mostly waits in a lounge, and Caroll sees her grandson from behind Plexiglas. Van is put on a waiting list for a heart when he is only a few days old. The last baby in intensive care who got a heart waited twenty-nine days. Even though no one is allowed to hold him, even though he is so sedated, Caroll falls deeply in love. She puts a photo of Van — with his head of pretty, thick black hair — on her office computer as a screen saver. Co-workers who see it react to the tubes and the swelling, "like it's the most awful thing they've ever seen, and I'm goin', 'Hunh?' Because I think it's the best picture," Caroll says. "I really think it is."

At church one Sunday, Pastor Keith and several servant-leaders ask people to come up for one-on-one special prayers. Caroll goes up there, to talk to Associate Pastor Ray Harmon, and she tells him she's sure God is going to make Van's heart okay, and that there's no more point in praying for a donor.

But a few days later, just after midnight on the Fourth of July (after sixty days of waiting), a heart does become available, from a baby who has died in Missouri. Outside the emergency room entrance, Caroll sees two doctors rush by, carrying the heart in a small red-and-white Igloo cooler.

Before the transplant, Van's hospital bills top $1.7 million. Michelle keeps worrying about the $2 million insurance cap. Caroll looks into what fundraising resources and support groups are out there for families of transplant babies. The next step would be the media alert, going public, opening the special bank account for donations from charities and faith-based action networks. This could bring on the KLTY radio wishes and everything else. But Michelle doesn't ask for that sort of attention, and it's not Caroll's style either. After the transplant, Caroll is allowed a peek, from afar, at Van. His chest is still open, she reports, covered, as if "by a sheet of Saran Wrap. You can see it beating in there. It's a heart."

Caroll adds a green organ-donor ribbon magnet to the back of her Taurus. She finds the tiniest flicker of optimism, or at least rational hope. "I said, 'Michelle, you've just got to not worry about this. God wouldn't let this happen, wouldn't let there be a heart and not have some other plan for how it's going to work out. You've got to believe.'" Caroll carefully thinks about what to say to her daughter, when to say it, and how to say it. She has always marveled at her daughter's independent streak. In Michelle's determination and in her reluctance to ask for help, Caroll sees herself. Thinking about how she sometimes keeps her own mother at a distance, Caroll worries about alienating

Michelle. They always got along best in the mall on Saturdays, shopping and talking, in the neutral arena of retail. But a hospital is different. Everything is different. Caroll broaches the issue of Michelle's faith. Michelle didn't get on the Pastor Keith bandwagon with her mother, and that's fine, but, "I told her, 'I don't know where you are with God, but you need him right now,'" Caroll says.

At one point, Van is hooked up to eighteen machines that work together to perform every single bodily function. Caroll has dozens of digital photos of what this looks like. In the center of those machines and wires and blinking red and green lights lies a baby in a clear plastic manger.

Here is where the miracle is supposed to come in, on cue.

I told you from the start: this is probably not that kind of Christmas book.

The baby gets the new heart.

The baby dies.

It happens on August 28, when Caroll has been thinking that maybe the worst is behind them. She knew that if Van ever left the hospital he faced a lifetime of serious health issues. She didn't care. She wanted him however he was.

As Van slowly fades away, his grandmother at last gets to hold him, for the first time. Caroll recalls for me that she stroked his skin and described for Van exactly where she believes he's going. She imagines him in a place where "it's a beautiful spring day, warm, and there's this big field with a babbling brook, and a big tree with a swing. I picture Van running around barefoot with his donor. And they play, and swing and stick their little feet in the water; they're little boys. God is sitting there watching them play . . . And I say, 'Lord, just love on Van for me, 'cause I can't.'"

Michelle and Joey don't want a funeral. They don't want a gravesite and they don't want a tombstone. Caroll asks if she can hold a small funeral service at church, or someplace else. But, she tells me, "Michelle said no. She said, 'We're going on with our lives.'"

Days go by, and there's some paperwork to fill out for the cremation. Caroll has suggested to Michelle that they should keep the ashes, in case they decide to do something with them, someday, years from now. It is only at the funeral home, Caroll says, that Michelle and Joey change their minds, after learning about common graves. She and Joey decide to pick out a simple urn and take the ashes home. Even if the urn stays in the back of a spare closet, Caroll is relieved. (Marissa says she saw it once in the TV armoire in the bedroom.)

"I don't want it like he was never here," she says, but she does what no other grandmother might have been able to do — she stays out of it and lets Michelle and Joey make all the decisions, the whole way. "Because I'm not the baby's mother, that's what it is." Still, "I wanted us to have a funeral, mainly for Marissa," Caroll says. "I didn't want Marissa to think that you die and you just disappear and nobody cares about you and you don't go anywhere."

Caroll feels spent, numb. She spends days cleaning out her garage and backyard, getting rid of things. Marissa's classmates and teachers have a prayer service for Van in the school chapel, and Caroll is able to bring photos and talk about her grief with everyone — at last. A woman at the school gives Caroll a bracelet with jade-green stones and a charm that has Van's initials with the dates he was born and died. After months of private Bible study, Caroll chances on a verse in the First Letter of Peter, chapter 5, verses 6 through 9, about humbling yourself to the Lord, and in due time, *He will restore you.*

"In due time," Caroll repeats.

She just has to wait for it.

Michelle tells Caroll something else: she's pregnant again.

Somewhere at Stonebriar Centre, on a Saturday in the early autumn, the jade charm bracelet falls off Caroll's wrist. She looks and looks for it, going back to every store she was in, asking every clerk, asking at Lost and Found, but it's gone.

20

Hot Topic

There's no recession in my world," Tammie Parnell tells me, coming up on Christmas of 2007. We're on ladders, decorating a fourteen-foot-tall tree together late one weeknight in the lobby of the private school that Blake and Emily attend. After Caroll's hard times, I'm oddly relieved to see Tammie. She takes me on an exhausting reprise of last Christmas's adventures in décor, and I'm happy to fluff garlands once more. Tammie has added another dozen Christmas clients this year for a total of forty-two. She's thought of taking her decorating business year-round, hiring herself out to stage empty homes for real-estate agents and such. She thinks of changing Two Elves with a Twist to something classier — "What do you think of 'Twist Design Group'?" she asks me. "Do you love it?"

Busier than ever, Tammie is still finding it difficult to actually stop and enjoy her favorite holiday. "I still just want the gift of time," she says. "I don't want the diamond Ebel watch — well I *do* want the diamond Ebel watch — but what I still want at Christmas is time."

One recent night, Tammie tells me, she was taking a long bath and Emily knocked on the door. She came in and tearfully told her mother that she knows — about Santa. Emily, now eleven, confessed that she eavesdropped on a phone call between Tammie and Tad as they were plotting her Christmas present—the big surprise was supposed to be that she's getting a cell phone. "Poor Emily," Tammie says. "That was so hard on her to be honest and tell me. It's one of those total rites of passage, you know? She said, 'I know it's you and Dad. There's just too many coincidences.'"

Tammie told Emily it's all right, that nothing changes, except that "you get to be part of the magic now. Help me keep Blake believing." Blake has lobbied Santa for a new PlayStation, which he won't get, but he will get a brand-new set of Callaway golf clubs.

On Christmas morning Santa will come for Emily anyhow, and he'll be there always for her, as long as there's a Tammie. Santa will leave clues all over the house for Emily to follow, until the last clue tells her to call a phone number. Her new cell phone will ring from deep within the tree.

I go back to the Hotties Christmas party with Tammie, this year at Helen's house, which is enormous. Once everyone has a few drinks, right before the gift swap, one of the Hotties distributes copies of a poem she wrote, tied to a frilly ornament—a gift for each of us. She works as a real-estate agent with her husband, and according to her rhymes, things aren't looking so good: "Hotties are HOT, we know this is so; but this Hottie's business has been very slow . . ." It goes on for several stanzas, asking her friends for customer referrals and reminding them to count their blessings.

A couple of months later, in early 2008, Tammie tells me it got worse. That Hottie and her husband have lost their custom-built, million-dollar house to the bank.

Between Christmas 2006 and Christmas 2008, some 55,000 homeowners in Dallas and Fort Worth (and the suburban cities clustered around them) have the foreclosure process begun on their mortgages, the highest number in decades. Some of the biggest jumps in foreclosures are seen in Collin County, in and around Frisco ("Frisclosure," *D* magazine jokes). A foreclosure in Texas can take a mere forty-one days to complete, about three months quicker than the national average. In 2007 these fore-closures include eight homes in Stonebriar Country Club Es-tates, where Tammie lives, and another sixty large homes in the fancy neighborhoods immediately around her. They include nearly fifty homes between Stonebriar Centre mall and Frisco Square. Twelve never-occupied townhouses in a new devel-opment a block south of Frisco Square had each been through at least one foreclosure by 2007. Only four homes in Hillcrest Estates — Jeff and Bridgette Trykoski's neighborhood — faced foreclosure in 2007, along with another four dozen homes in adjacent neighborhoods. On the former farmlands due west of the Dallas North Tollway there are thirty-eight foreclosures, all in new subdivisions. In and around Caroll Cavazos's neighbor-hood in the Colony, there are twenty-eight foreclosures started and another twenty or so in neighborhoods nearby. By the end of 2008, the foreclosure rates in some of these areas have again doubled.

Frisco has gained about 10,000 new dreamers of the dream in the last eighteen months, more families in search of its top-rated schools, big houses, and convenient shopping. The num-bers vary and it is impossible to know the exact day and per-son, but someone became Frisco's 100,000th resident in 2008. Symbolically, that person should be a baby, probably named Delaynee or Kaydee, Tyson or Corbin (or as likely Josefina or Daniela, Refugio or even Jesús). The eventual "build-out" popu-lation estimate by city officials predicts another 125,000 or so newcomers in another ten or fifteen Christmases from now. The population slows with the economic stutter, but it is still enough to keep the city on a number of lists of the fastest-growing (and highest-income) suburban cities in the nation. An analysis of FBI crime statistics in 385 American cities names Frisco as the sixteenth-safest city in America. Even now, the people I meet remark how safe they feel here, from a variety of real and imagined terrors. Being here keeps them at a prudent remove.

Sometimes I get the feeling they are too far removed. Voter turnout for city elections is abysmally low: a new mayor, Maher Maso, is elected in 2008 with 5,358 votes to the 1,601 votes won by his opponent, Matt Lafata. The city opens a new history and heritage museum filled with farm implements and antique furniture and continues with plans for a railroad museum and a "Grand Park" of carefully selected and ecologically appropriate nature. Yet the general citizenry's awareness of local issues and history seems neither here nor there. Often Frisco is a world of cheerful zombies, who move among containers (home, mall, car) as if in a narcissistic family-first and property-values haze.

When I guest-judge the interview portion of the Mrs. Frisco United States pageant in 2007, only one of the nine candidates is able to name the mayor or any of Frisco's city councilors. The pageant isn't a pageant; it's held on a Saturday morning in an empty office conference room in Frisco Square. Our Mrs. Frisco is pretty and nice, with a platform focused on her own child's struggle with a rare heart disease, but she does not go on to win the state crown.

Circuit City
(SOME ENDINGS)

B EFORE THE BLACK FRIDAY dawn, on November 23, 2007, I pull into a different Frisco strip mall this time, a lot like last year's but across the road. Caroll's son, Ryan, has spent the night with some friends, camped in front of Circuit City instead of Best Buy, claiming a spot and pitching a tent near the front of a line of people that stretches well past the Bed Bath & Beyond. In the summer, Ryan had moved out to live in a friend's apartment and continue working at Best Buy, but he's getting ready to move back home with Caroll. He's taken on a second job, parking cars at the mall's valet station. He's still trying to figure out a way to get back to Oral Roberts University.

Caroll and Marissa stand and shiver in the brisk air. Caroll has decided to get her mother a new TV this Christmas, splitting the cost with one of her sisters again. The TV is just like the one Caroll has — a fifty-inch Samsung flat-screen, which Circuit City has on sale for $799.99, down from the regular price of $1,199.99. It's the only thing they've come for. She has still budgeted about $300 each for Michelle and Joey, for Ryan, and for Marissa. Michelle and Joey's new baby is due in May, but they haven't given Caroll any wish-list items, since they seem to no longer wish for much. Ryan told Caroll he wants the Blu-Ray boxed set of *Planet Earth,* the Discovery Channel documentary.

"I'm rethinking how we do Christmas," Caroll says. "We need to change. If nobody wants anything, then what's the point?" What's the point of going with Ryan to the store so he can buy what he wants and have Caroll pay for it? Where's the joy? "It just becomes this thing where I'm asking them what stuff to buy, and they don't even want anything."

Marissa blurts out an objection to this, hopping up and down: "I do, I do! I just don't know what!"

The doors open and the crowd surges forward to get into Circuit City, but they let people in only a few at a time. We stand in the parking lot and watch as Ryan finally goes in, and then we wait even longer for him to call Caroll's cell phone with the signal that he has the TV and is ready to pay with Caroll's Bank of America card. As we wait, a heavyset woman unhappily huffs out of the store and announces loudly to anyone within earshot: "Don't sweat it, people — nothing's left."

Ryan calls. Caroll and Marissa go toward the door, to push in with the crowd, but this time I'm staying put.

Jeff Trykoski adds thousands more lights to Frisco Square's Christmas displays of 2007 and 2008, stringing 8,000 feet of coated galvanized aircraft cable across the square in forty-eight neat rows to create a canopy of light. With a bigger budget, Jeff orders six improved snow machines, costing a few grand each. He also orders fifty-five-gallon drums of deionized water from a cosmetics manufacturer in Oregon, to make better snow.

The local news stations all feature stories about Jeff. He and Bridgette are also one of three families featured on a TV documentary series called *Magnificent Obsessions,* which had been shot a year prior, in November 2006, and is being shown on a lifestyle channel available to Dish Network subscribers. Jeff's obsession is described alongside that of a Muskogee, Oklahoma, man who has one of the largest public displays of air-blown holiday characters, and a Brooklyn man who has a computerized light show on his row house. Each of these guys gives entirely bland answers to the question of why they do it, other than they love Christmas, and they love giving back to people, and isn't that reason enough? (And it's not as if Mike Wallace was doing the asking.) I've been around Jeff long enough to understand that he simply loves the task of doing it, the business, the service. Greg Trykoski says it's the one time of year his brother gets to be a rock star.

Jeff's magnificent obsession goes on, perhaps past the limit where I (or Bridgette, or his parents, or even his brother) can ever go, but we go anyhow. His plan now is to cut out the middleman of retail wherever possible. In the Christmas-lights world, the craze is on to convert displays of incandescent bulbs to all-LED bulbs, which give off a more energy-efficient glow. There are two camps when it comes to "going LED": some prefer the bright incandescence of Christmas past, and some prefer the eco-wise shine of Christmas future. Jeff is fascinated by LEDs, but at a dollar a bulb, an LED display on a Trykoski scale would run many more thousands of dollars than he's already spent. He does the cost analysis anyhow and concludes that he should get in on the capitalism. He winds up ordering an entire boxcar-sized shipping container of LED lights in the spring of 2008 through a vendor that works directly with a factory in China.

The LEDs are made in Dongguan, in Guangdong province, some 100 miles northwest of Hong Kong and 7,230 miles from the Trykoski house. The factory employs as many as 1,200 Chinese workers in the spring and summer, as Christmas is assembled. When an earthquake hits Sichuan province in May 2008, killing more than 10,000 people, Jeff sends his brother Greg a quick text: *Earthquake in China. Lights OK.* (The quake's epicenter in central China was more than 1,000 miles from the coastal factory.)

Twenty-seven thousand sets of individually packed LED light strings are boxed and loaded into a blue Changzhou Xinhuachang International Containers Co. boxcar container addressed to Jeff. He has naturally done the math: it is 1,365,000 bulbs, or eighty-five miles of lights end to end. Each cardboard box in the shipping container weighs between six and nine kilograms (thirteen to nineteen pounds) and holds twenty-four sets of lights. The container leaves Hong Kong on a transoceanic super-freighter on June 15, one of hundreds of similar containers that are on board, most carrying some form of merchandise. Depending on the price of oil, such sea voyages now cost nearly $5 million each way. From the harbor at Long Beach, California, Jeff's container travels by rail and truck to Texas. Before the ship even leaves China, Jeff has presold LEDs in bulk orders to customers as far away as Minnesota. A third of his shipment is sold to a resort hotel, and almost as much goes to a Baptist church for its Nativity pageant.

It's a sizzling Friday morning in mid-July at the U-Store-It facility next to Jeff's neighborhood, where we're all waiting to open the Chinese container. Jeff and Bridgette have recruited twenty guys to come help unload the LEDs and sort them according to type — "mini ice multi on green wire," "C6 faceted pinecone," "pure white on green wire," "red on green wire," and so on. The container bears Chinese characters stenciled on its doors and is pulled by a semi with Tennessee plates and temporary Texas tags, driven by a Latino. We had talked for weeks about what this moment would be like, when the plastic bolt is hacked off and the doors swing open. Greg had made jokes about the container being filled with dead Chinese immigrants.

The doors swing open.

It takes an hour to unload and then another hour to sort. Bridgette starts yelling at Jeff, who has criticized her inventory method. "What, Jeff?! I'm trying to do that, dammit!" she yells. "Fine, if you don't like the way I'm doing it!"

"One hour, forty-two minutes," Greg says, pretending to check his watch against his sister-in-law's temper. "About right."

"My wife could yell at me, too," Bryce Kindla says. "But it wouldn't change the fact that *I have all these lights.*"

The humiliated though unapologetic titans of Wall Street refer again and again in late 2008 to a "perfect storm" of worldwide conditions in the consumer credit and investment markets that "nobody saw coming" (when, in fact, plenty of people saw it coming). *Bailout* becomes a national buzzword, a trillion-dollar rescue at first that becomes pleas for even more; everyone seems to need immediate aid, from the lowliest shopper to the overextended homeowner to Bear Stearns to Freddie and Fannie Mac to the Big Three automakers. Our economic woes verge on calling out for familiar magic, a "Dear Santa" letter gone unanswered, begging the Federal Reserve Board to lower the interest rate all the way to zero. (Which it does, nine days before the Christmas of 2008, while Freddie and Fannie announce a holiday moratorium on December/January foreclosure proceedings, like something out of a black-and-white movie. Still hunting for the Christmas metaphor in almost anything, I notice that Ben Bernanke, the Fed chairman, has a white beard; now all he needs is a red suit.)

There is no short supply of economists pointing out the obvious — that we'd believed for too long in the myth of easy credit, cheap consumer sprees, outsourced labor, and imbalanced trade. "It seemed too good to be true, and it was," a Harvard economist, Kenneth Rogoff, told *USA Today,* leading up to one disastrous stock dive and investment-banking collapse after another. "Absolutely [this] is payback. The best-case scenario is we have a long but mild recession — and that's the best-case scenario." One of the earliest responses to an American consumption crisis is congressional approval of a $153 billion economic "stimulus package," a rebate check of anywhere from $300 to $1,200 mailed to most taxpayers. We are then supposed to take these checks straight to the mall and spend like it's Christmas. By June 2008, when most of the checks had been cashed, sales data from the nation's chain stores show the barest increase — half a percentage point. Americans did the wrong thing with the money. They used to it pay off their credit cards, or worse, they saved it.

"What Recession? Frisco Holds Its Own" is the headline on a *Frisco Style* magazine story that same spring, making a case that the nation's woes have largely passed over their paradise. Frisco's total sales tax revenues, on what was now close to 7 million square feet of retail space, amounts to some $40 million in 2008, an increase of 112 percent in six years. Another 900,000 square feet of retail space, much of it in brand-new strip malls, will sit unoccupied in the first half of 2009. Frisco's officials welcome the slowdown as a chance to breathe and to revisit the master plan.

Not everyone agrees with the rosiest outlooks. When delivering her annual report in June 2008, Jill Cumnock, the executive director of Frisco Family Services Center, noted that 300 more families are expected to need her center's assistance by Christmas. Gifts needed for children in the Angel Tree program increased from about 600 in 2007 to more than 1,000 a year later. The food bank now runs chronically low. Cumnock has seen a trend among new clients: Many of them, she said in her report, once worked in the developing, building, or selling of real-estate properties. They had worked at mortgage banks. They had worked in high tech. Many had simply been caught up in the notion that there would always be more.

All along, in good times and not-so-good, people keep pointing me toward "better" Christmases than the ones they are presently having, recommending examples of more "amazing" holiday moments, which always seem to be happening in other towns to other people. They suggest the lakeside lights at a festival in Natchitoches, a small Louisiana town "where they made *Steel Magnolias.*" They suggest villages and quaint-sounding burgs in Minnesota and Vermont. Some feel it negligent to write a book about extreme holidays and skip Manhattan, the birthplace of American Christmas, leaving out the Rockettes, the Barneys windows, the Macy's Thanksgiving Day parade, and the tree lighting at Rockefeller Plaza. Someone else talks about a multimillion-dollar Christmas pageant I need to see in Florida, and another mentions one in Pennsylvania.

They say if I really want to see beautiful Christmas lights on lovely homes, I should take a nighttime drive through Highland Park, an old-money Dallas neighborhood, or through the castle homes of Deer Creek in Plano. In the malls, people waiting in line to see one Santa tell me about Santas who are better (and more beloved) at other malls. People point me to "better" people, too, by which they mean people having nobler, more storybook Noëls: compassionate doctors, returning servicemen, members of a family who stage an elaborate Christmas scavenger hunt for one another. What about the big party in Dallas where Neiman Marcus unveils its phantasmagorically luxurious Christmas "Book" with the $48,000 pavé diamond cuff, the $5,000 Airstream trailers for dogs, and the $1.7 million rides on the Virgin spaceship?

I followed all these leads, but in the end, there was one thing that gave Christmas its best reason of all, and it's something I do not have, and never will: a child.

It is true that nearly everything about the American Christmas defies logic; at the same time it is made acceptable and infinitely wonderful by the simple fact that it delights children — and therefore delights the child within almost any adult. When it comes to Christmas in Frisco, people try to replace what they've lost — tradition, magic, the past — by finding it again in their children. That was the ending to every Christmas story, and the beginning of it, too.

On another trip to Frisco, in the summer of 2008, I find Caroll at the mall, with Michelle and Marissa, on their usual Saturday outing. It's one of the last times I see Caroll. Her new grandson's name is Lincoln. He's a robust little guy with Michelin Man arms, and Caroll pushes him around in his sport-utility stroller. After a late lunch at the California Pizza Kitchen, Michelle goes into the Disney Store and comes out with the baby's first pair of sunglasses. Caroll reaches into her purse for her camera and snaps digital pictures of little Mr. Cool. She's happy. Right now. Here is where she can once again put all her hope.

> Jeeves was in the sitting-room messing about with holly, for we would soon be having Christmas at our throats, and he is always a stickler for doing the right thing.
>
> — P. G. WODEHOUSE, "Jeeves and the Greasy Bird"

Wodehouse said Christmas was at our throats, and some say Noël Coward used to write the same thing in his annual holiday card: *Christmas is again at our throats.* Archie Bunker said it in an old episode of *All in the Family:* "Ah, jeez, Christmas is at our throats again."

Suppose we did find a way to scale back on Christmas and

therefore breathe easily? If enough American consumers cut their holiday spending, even by half, the consequences to our way of life would be a disastrous chain reaction of job losses — in the retail sector, but also in shipping and handling and other service sectors, to say nothing of the effects on global manufacturing, marketing, and all forms of commercial media. There would be a final death plunge in the stock market. While we pine for simpler Christmases of handmade toys and ornaments, the effect of such a holiday could truly transport us back into the very world of the Depression-era American Girl dolls we so envy for their make-do sincerity. You don't have to study Christmas math very long to realize we can never go back.

If there were such a thing as the Grinch, he wouldn't sneak in at night and steal toys. He'd be vigorously brooding (and blogging) about the end of fossil fuels and lecturing people about their plastic lifestyles. (He'd be a little bit like the Reverend Billy Talen, that New York performance artist in the bleached-blond pompadour and white clerical robe who leads his Church of Stop Shopping toward a holy "shopocalypse," traveling the country to stage anticonsumer religious revivals at big-box stores and malls.) The Grinch would be asking how many more years this lifestyle can stand: How can everyone's daughter be a princess? How can everyone be entitled to ever-doubling values in their homes? He'd show you charts of what happens once the oil runs out, midcentury. How long before we could no longer afford to ship Christmas to wherever the richest consumers live? How long before the malls became haunted, empty husks? He'd be ranting at the top of his green lungs and no one would hear him.

Christmas is at our throats again. I go back one last time in early December 2008, to take in the abnormality of a season unhinged: Shopocalypse Now.

Caroll and Marissa report that they arose early as usual for Black Friday on November 28, but it isn't the same. That first year we met at Best Buy. Last year it was Circuit City. This year, "I have just as much money saved up for Christmas as I have every year," Caroll says, "but you can just tell, it's different out there. Nobody wants anything big this year, so why buy it? Why spend it?"

Caroll still likes the tradition. Even though it's a wet and unappealing morning, she and Marissa traipse to J. C. Penney to get their free Mickey Mouse snow globe — their eighth year of the collection. They head to Kohl's to hit a sale on two digital picture frames, which Caroll is giving to her sisters. They go to Toys R Us to fulfill the request of one of their Frisco Family Services Angel Tree children — "Girl, 6," who asks for a baby doll. Caroll falls completely for a set of twin dolls that play peekaboo and sing patty-cake rhymes. But they're black. She looks on the shelf and finds another pair, a blond and a brunette, but these speak Spanish. "Now, I'm goin', either one of these could be appropriate, but you just don't know, you can't be sure," she tells me. A sales clerk hunts in the back and returns with the perfect compromise, the American dream — creamy, possibly ethnic babies who speak English.

Ryan is now almost twenty-two and has met a nice girl; she works at the Petco. It's a strip-mall romance. He has to work the opening shift at Best Buy this Black Friday, so instead of camping and waiting with the hordes, he's now part of the team trying to keep the peace when the crowd comes streaming in. (In Long Island, a man is trampled to death when the doors open at a Wal-Mart.) These are Ryan's waning days at Best Buy. He's decided to go to Oklahoma State University in January and study industrial engineering and management. Come Christmas, it will be Caroll who finally gets the big presents: Ryan gives her a GPS device for her car. (Never get lost.) Michelle gives her mother a zebra-patterned Dooney & Bourke handbag. "I was blown away," Caroll will tell me. "My kids really did for me."

Yet the shopping we do is still never enough. The Saturday after Black Friday, the National Retail Federation issues its routine perky findings: Sales are up! More people than ever — 172 million, they tell us — got out there and shopped! But on Monday, more sobering news takes hold. The U.S. economy is now officially declared to be in recession, and it has been since Christmastime 2007. Unemployment — nearly 2 million jobs are lost in 2008 — hasn't been this high since the early 1980s. The consumer confidence level is lower than it has been since they first started measuring it. There is also new evidence of a doomsday scenario: shoppers have stopped using their credit cards. "Christmas is over. It's dead. It's going to be horrendous and it's going to be negative," remarks one analyst, Howard Davidowitz, to the *Star-Ledger* in New Jersey at the beginning of the 2008 holiday season. "The consumer is murdered. There's no way back."

It's heebie-jeebie vibes in the homogeneous habitat. KB Toys files for bankruptcy for the second time and initiates a going-out-of-business sale at all its stores two weeks before Christmas. Circuit City will go under a couple of weeks after Christmas. Club Libby Lu is shutting down all its tween boutiques. (Farewell, belly-shirted glitter imps!) Sharper Image is gone, the Bennigan's restaurants are shuttered, Linens 'n Things is kaput. It's a long series of omens — the end of *things* 'n things.

I cruise among Stonebriar's Christmas shoppers and we exchange grim glances. It feels off. It's our communal going-out-of-business sale. It's beginning to feel a *not* like Christmas. "When the dust settles, it will still be the weakest holiday season back to 1970," the International Council of Shopping Centers'

Michael P. Niemira tells the *Miami Herald*. "It's too late to do much to save the holiday season." (In January 2009, the numbers are out: for the first time anyone can remember, Christmas retail sales declined in America, by nearly 3 percent. Three months after that, General Growth Properties, which operates Stonebriar and 200 other shopping malls, files for Chapter 11 bankruptcy, seeking protection from billions of dollars in unpaid debt.)

I'm in Stonebriar several times before I notice the real change: not even half of its usual Christmas decorations have been put up this year. The genuinely bearded Santa Claus, who had sat in his green velvet chair every year since the mall had opened in 2000, is gone, replaced by another genuinely bearded Santa. There is great angst about the new Santa. He is "too grumpy," according to online comments on a website for Frisco residents, and he has an unfortunate feathery helmet of hair, looking much like an early-1980s TV anchorwoman. He is loathed.

Even the songs in American shopping malls sound somehow wrong this year. Mod club DJs have gotten their hands on the standards and remixed them into dirges. Eartha Kitt's "Santa Baby" sounds tranquilized, slowed down with a hip-hoppish lounge backbeat: "An. Outer. Space. Convertible. Too. Bright. Blue." Perry Como sounds doped up with her: "To. Face. Unafraid. The. Plans. That. We. Made . . ."

Another Chick-fil-A opens in Frisco, and the people arrive to camp together overnight in the chill, to be first to claim books of free combo meal coupons at the celebratory unveiling. When I see this now, I get the unshakable sensation that it's some dress rehearsal for real calamity, when we'll all be living in strip-mall parking lots. The wonder in them is the pessimist in me.

* * *

The Brothers Trykoski make themselves sick of Christmas one more time for the greater good of a winter wonderland of lights, listening over and over to the songs they and Bridgette have selected for Frisco Square, which now includes two additional buildings and another 30,000 bulbs to synchronize to the music. It's "Carol of the Bells" by Mannheim Steamroller, Frank Sinatra's version of "Jingle Bells," "Little St. Nick" by the Beach Boys, and "Feliz Navidad" by Céline Dion.

At the Merry Main Street festival, the mayor of Frisco takes the stage and asks the crowd to help him count down to the lighting of the town tree in the plaza center. "Five," the mayor starts, and the crowd joins in. "Four! Three! Two! One!"

Nothing.

We all boo.

"Okay, okay," the mayor says. "That just means we have to do it louder! Ready? Five! Four!"

Three! Two! One!

Still dark. The crowd boos again and then, a couple of agonizing seconds later, the tree comes on.

The tree happens to have on it the only lights here *not* controlled by Jeff Trykoski. The city is supposed to flip that switch itself. The mayor immediately gets out his phone and texts Jeff: *Next year, you're doing the tree.*

A producer from *Live with Regis and Kelly* calls a week later, asking to air a video of the Trykoski house, the original YouTube clip that by now has received nearly 2 million hits. Jeff agrees, but he has become a little circumspect about small-time fame. "A lot of people get into this because they want to be the next Carson Williams, or the next me," he says. "They want media attention. They're not doing it for the right reasons."

Christmas lights — creating displays for the square, selling LEDs — has been a boon for Jeff and Bridgette in an economy where so much else has been cut back. He'd used the extra cash

to finish his MBA, Bridgette got a new Lexus, they went to Can-
cún, and they're installing a pool in their backyard.

Jeff is thinking about running for city council someday. He
wants people to see Frisco the way he sees it, as something more
than just a strip-mall town full of people who don't care enough
about where they live. He's not doing the lights to save the retail
prospects of Frisco Square. He tells me he wants his town to be
a real place, with a real Christmas that people will remember
for decades to come. "You only get one shot at this," he says.

"A shot at what?" I ask him.

"Making a city," he says.

To face unafraid the plans that we made.

The stress of Christmas lights had seemed to be getting the
best of Jeff when I saw him in summer, when the container arrived from China. But he seems happier now and less . . . wired.
The lights will either work or they won't, but he's not going to
kill himself trying to make them flawless. Greg programs the
display on the Trykoski house to dance to the mellow sounds of
Jimmy Buffett singing "Mele Kalikimaka," the Hawaiian Christmas song. Bridgette tells me she and Jeff are doing everything
they can to just have a quiet, relaxed Christmas. "I keep reminding Jeff how boring his life would be without me," she says.
"Without me, he'd just be a lonely rich guy."

The geek in Jeff still shines through. At a party at Greg and
his girlfriend Christine's place after Merry Main Street, Jeff suddenly announces, apropos of nothing: "The traffic at our house
is going to be insane this year. People are going to have to wait
forever. It's going to be bigger than ever."

"Jeff, there's this little concept called humility," Greg says. "It
would be a good thing to learn. You don't just say, 'You won't
even be able to get into the cul-de-sac tomorrow night.' You say,
'Yeah, we have some lights on the house, and some people might
drive by and see them, it's cool, whatever.' It's like me saying, 'I

have a huge penis' . . . You can put that in the book, by the way. *I have a huge penis . . .*"

In the end, it will be up to Tammie Parnell to save the American economy. Having seen her beloved new soul mate — the vice-presidential candidate and Christ-centered Alaska Governor Sarah Palin, who seems like a total Hottie — vanquished in the November election, Tammie is more determined than ever to make Christmas happen for us all, while making it happen for her bottom line. In addition to decorating a couple of dozen houses this season (down from last year's forty-two; she admits some of the clients have cut back their spending), she's blown the bazaar circuit wide open. Let the Fed and Treasury figure out the macro economy; Tammie has gone micro.

She went to all the market shows this summer and loaded up on knockoff and one-of-a-kind leather handbags and jewelry from China, along with heaps of pashmina shawls and fancy scarves, in addition to twice as much Christmas décor. She's set up sales tables at every church, Junior League, and school bazaar. She goes to people's houses and puts on sales parties. She even goes to a Jewish woman's house in Dallas to sell holiday finery, willing to acquiesce to the hostess and weed out any item with a cross on it. "You gotta go where that dollar is," she says. "I'm not worried, really. Some people always have money." Still, there is a lesson about greed to be had, and Tammie is counting her blessings. "Pigs get fat and hogs get slaughtered," she tells me one more time, one of her favorite mottoes. (I've yet to fully comprehend that one. Fat or slaughtered, neither sounds so great.)

The last time I see Tammie is in the lobby of an office tower near the Dallas North Tollway, down near the Galleria shopping mall. This is what she does now — sets up a few tables in office lobbies and sunlit atriums, out there with the badly decorated fake Christmas trees and the piped-in Muzak carols. Gals come down from their offices to look at Tammie's display of purses and pashminas, the red-and-white candy cane dishware, and plaques that say "Be naughty — save Santa the trip" and "*Three* wise men!? Are you serious?" and "I'll get my elves right on that." Tammie likes cash, but she'll take your check, and now, with her little wireless gizmo, she can take any credit card but American Express. "I'm doin' absolutely phenomenal in these office buildings," she says.

Well, there was one slow day. It was in a building emblazoned with the logo of Lending.com, the headquarters of an online mortgage broker. Tammie complained when she got home that night: it seemed as though nobody was in that building, and the people who were there didn't want to buy any of her stuff.

"Tad and I got in this big fight," she tells me. "He was saying, 'Lending.com?! Tammie, those people are all about to lose their jobs! They're not going to buy your Christmas stuff!'" She understood where Tad was coming from. He works in retail menswear, "and that's just awful right now. Tad says I'm living in a bubble."

Tammie resolves to spend only $500 each on Blake and Emily this year. "I was telling them this morning, 'You are pampered. You've got it too easy. You have everything you want, you get driven everywhere.' I mean, $3,000 to sign up for team volleyball this year. There's this whole world out there they don't know about." Emily is twelve and still wants an Apple MacBook laptop, which starts at $1,299. "Nunh-unh!" Tammie says. "She can deal with a Dell." (Guess what? Emily got the Mac. Once in the Apple Store, Tammie just can't resist.) And what about Blake, now nine, who is still clinging to his belief in Santa? "Oh, easy this year," she says. "A Browning twenty-gauge shotgun. Can you believe it?"

Tad's bubble comment has clearly gotten to Tammie. She

tells all her customers the story: "My husband says I'm living in a bubble."

Tammie and I say goodbye in the parking lot.

"Do you think I live in a bubble?" she asks.

I drive faster than I ever have, getting on the tollway, merging onto the George Bush, and taking a straight shot to D/FW. At the Enterprise Rent-A-Car lot I toss my keys to the attendant, catch the shuttle bus to an American Airlines ticket counter, and sail through security, hearing the voice of Darlene Love coming out of the Chili's in Terminal C, and then over and over in my head: *Baby, please come home.*

Yes, is what I told her.

Yes, Tammie, you live in a complete and total bubble.

I suggested we call it a snow globe.

She liked that.

As my plane takes off, I put names to what I know of what I see across the luminescent grid — suburbs, malls, tollways, interstates. Quickly it disappears behind the clouds, my sparkling snow globe filled with all that synthetic cheer. Christmas is at our throats again, but I can feel it letting go of mine.

ACKNOWLEDGMENTS

Would you be willing to let a stranger spend Christmas with your family? While he takes notes? Even when he asks how much you spent on everything? Would you let him come with you to work, school, the mall, grocery store, church, parties, dinners out, your kids' volleyball games? And what if he stuck around a year longer than he said he would?

The people written about in this book let me see them as they are, while answering almost all of my questions. They fed me at their tables, prayed for me at their churches, promptly returned my calls, and asked me some smart questions of their own. They treated me as a friend, which, as any journalist knows, put us in tricky territory. In return, I can only hope I've portrayed them as accurately as possible, with my respect and admiration. So, to Jeff and Bridgette Trykoski and the Trykoski and Iraggi families; to Caroll Cavazos and Marissa, Ryan, Michelle, and Joey; and to Tammie, Tad, Emily, and Blake Parnell — thank you all. I hope we hear from one another for many Decembers to come.

Dozens of people in Frisco and Plano — among them ministers, family counselors, social workers, business owners, wheeler-dealers, and Junior Leaguers — gave me a Texas welcome and pointed me in many intriguing directions. Frisco's city officials were helpful, including then-mayor Mike Simpson, city manager George Purefoy, mayor emeritus Bob Warren, and

Jim Gandy and his colleagues at the Frisco Economic Development Corporation. Dana Baird-Hanks, the city's communications director, provided lots of good leads early on. The folks at Stonebriar Centre — from the management office to the mall cops to even the bartenders at the Cheesecake Factory — put up with what must have seemed a creepy amount of time for one man to spend in their world. Jill Cumnock and her staff at Frisco Family Services Center were most kind, just when they were most busy helping the people who truly needed them. The anonymous denizens of Frisco Online's "Talk About Town" and "Frisco Focus" forums unwittingly provided shrewd insight, comic relief, and bafflingly good gossip. In many ways, it's a wonderful life out there, and I salute the people who made it and live it.

My thanks go also to Ben and Shelly Beckelman and their daughters; to Denise, Hollis, and Danielle Matise; to Elizabeth Brzeski; and to other families and individuals with whom I visited and about whom I took many hours of notes before settling on other story lines. I'd also like to thank Brett Jensen, Sherrill Jackson, Courtney Perry, Gary Cathey at Daryan Display, the members of the Texas Christmas Lights Club, Beth Robinson, Randy Moir, Robert Medigovich and the recyclable-waste crew at CWD, Tony Felker, Matt and Erika Lafata, Celebration Covenant Church, Prestonwood Baptist Church, the "Gold Hotties," Belle Marie Demarest at Holidaze & Gifts, and Janis Jackson and Marian Chadwick at Extravagant Necessities.

Laura McCall Froelich (aka "Derba"), to whom this book is partly dedicated, has been a treasured friend since we first met in high school, and she looked at every draft of this project with a keen, north-of-the-LBJ eye. Laura and her husband, Dan (and their kids, Bobby and Annie, and their dog, Mujibar), let their Dallas home become my safe haven many times. As a bonus, I frequently got to enjoy the wit and inquiring mind of Laura's mother, Winnie "Wee-Wee" McCall.

Other friends in Dallas provided excellent company when I needed a break, especially Tim Flannery, Mike Thompson, and Kelly Smith. My beloved uncle, Louis Schneider, and his partner, John Johnson, treated me to many fun excursions in Big D, including frequent breakfasts at Cindi's and lunches at Hunky's.

Heather Schroder at International Creative Management was always ready to help with smart advice, and if she weren't already the perfect agent, I'd want her for an editor.

But this book is already lucky to have had two remarkable editors. George Hodgman believed in the idea from the very first time I brought it up at lunch in 2005 and talked me through all the field reporting and my embarrassingly rough draft. When George left our first publisher (the very fine Henry Holt) for a new job at Houghton Mifflin Harcourt, he brought me (and my book contract) with him, which he did not have to do. Although George and I did not get to the last page together, I thank him deeply for his work and friendship.

I am so grateful to Andrea Schulz, Houghton's editor in chief, for making room for this book on her busy desk. She read the manuscript in a week, figured out exactly what it needed (and what it did not), and articulated her thoughts in one of the clearest memos I've ever received. My thanks also go to the many people in Houghton's editorial, production, and marketing departments who gave this book so much enthusiastic attention. And to Barbara Wood, who left a nicely wrapped copyedit under the tree.

The generous people who run the *Washington Post* gave me fourteen months off to work on this book, during a less than ideal time for any newspaper employee to step out of the frenzy that is remaking American media. My *Post* editors — Deborah Heard, Steve Reiss, Henry Allen, and Ann Gerhart — understood that reporters don't exactly return from a project like this in peak form, and I'm sure I stretched their patience as I tried to finish the book and file stories. My compadres in the Style section were similarly supportive, especially Leslie Yazel, Paul Farhi, and Bill Booth.

Judy Coode is a good ear (and always has been), when the question "How's it going?" predictably sends me into self-deprecating monologues. My mother, Joann Stuever, and the rest of my family put up with a fair amount of that as well and helped me remember some details about our Christmases together, way back when. I am grateful to those friends who volunteered to read this book in various forms and got back to me with critical notes and observations. They are Janet Duckworth, Jessie Milligan, Linda Perlstein, Kelly Brewer, Laura Trujillo Faherty, Elaine Beebe, and, to my everlasting relief, the unrivaled Pat "the Perfect" Myers.

Finally, I want to say that I try very hard not to take Michael Wichita's love for granted. He stayed home while I went out in search of the American way of Yule and didn't seem to mind when I came back slightly schizoid. We lost a lot of time together because of this book, but what I like in Christmas songs are the parts where they sing about *next* year, and the happy Christmases that will come. My big hope now (and always) is that those songs are about Michael and me.

Oh, and PS: I'd like to acknowledge the global economy, especially the credit and retail sectors, which fell apart between 2006 and 2008 and thereby made profligate Christmas shopping seem all the more interesting and a bit more inane. Here's to you, capitalism.

SOURCES, BIBLIOGRAPHY, AND SOME
OTHER STOCKING STUFFERS

If someone starts telling you with absolute authority about the origin of the candy cane, or where the legend of Santa Claus came from, or how Christmas traditions go all the way back to Bethlehem, you should consult other sources if you simply can't take it on faith. Start with these:

Marling, Karal Ann. *Merry Christmas! Celebrating America's Greatest Holiday.* Cambridge: Harvard University Press, 2000.

Nissenbaum, Stephen. *The Battle for Christmas: A Cultural History of America's Most Cherished Holiday.* New York: Alfred A. Knopf, 1996.

Restad, Penne L. *Christmas in America: A History.* New York: Oxford University Press, 1995.

In the tangled heap of dubious and overly sentimental writing about the holiday, I was happy to have Marling, Nissenbaum, and Restad ready on a table next to my desk. I reference Marling on collectible snow villages in chapter 9, and Nissenbaum in chapters 7 and 8, regarding the early evolution of the American Christmas and New York charity events in the 1890s. Restad's and Marling's summaries of early use of electric Christmas lights came in handy in chapter 4.

Similarly, there is a lot of writing on the religious nature of Christmas and its theological meaning and importance. Like so many others, I looked to:

Brown, Raymond E. *The Birth of the Messiah: A Commentary on the Infancy Narratives in the Gospels of Matthew and Luke.* New York: Doubleday, 1993.

Here are two other takes on Nativity scholarship that I recommend:

Horsely, Richard A. *The Liberation of Christmas: The Infancy Narratives in Social Context.* New York: Crossroad, 1988.
Spong, John Shelby. *Born of a Woman.* New York: Harper-Collins, 1992.

When it comes to the modern settlement of North Texas, there are many tall tales and creation myths, especially about the communities beyond Dallas. As Frisco neared its centennial in 2002, some of its citizens hired a *Dallas Morning News* reporter to get their stories straight, and I tip my hat to his thoroughness:

Quinn, Steve. *Frisco: The First One Hundred Years.* Frisco: Heritage Association of Frisco, 2002.

I also found these works helpful, as a glimpse of rural life in Collin County before strip malls:

Bowyer, John Wilson, and Claude Harrison Thurman, eds. *The Annals of Elder Horn: Early Life in the Southwest.* New York: Richard R. Smith, 1930.
Clark, Adelle Rogers. *Lebanon on the Preston: A Casual Biography of a Backland Village.* Wolfe City, TX: Henington Publishing Co., 1959.

Connor, Seymour V. *The Peters Colony of Texas: A History and Biographical Sketches of the Early Settlers.* Austin: Texas State Historical Association, 1959.

Warren, Robert M. *Frisco—Now and Then.* Baltimore: Gateway Press, 2004.

More Sources . . .

. . . On Christmas Spending

The National Retail Federation in Washington, D.C., and the International Council of Shopping Centers in New York calculate American spending during the holiday season in different ways, using different metrics and research. The U.S. Department of Commerce numbers are different from those, and yet supplementary. In seasons that are markedly robust or sluggish, these various data lead economists to the same general conclusions about sales growth or decline.

Like most other journalists, I relied — perhaps too much — on what the NRF says. (You see them attributed in almost all Christmas-related business stories in December, the same way you'll frequently find the ICSC's chief economist, Michael P. Niemira, quoted about how the holiday shopping season is going.) Where I've used numbers, they usually come directly from those groups — especially the NRF's tally of the nearly half-trillion dollars spent on Christmas gifts in 2006 ($456.2 billion). That number remained almost flat in 2007 and declined, in 2008, to $447.5 billion.

My reference in "Best Buy (A Prologue)" to a single Christmas costing more than our current wars (up through fiscal year 2006) comes from "The Cost of Iraq, Afghanistan, and Other Global War on Terror Operations Since 9/11," prepared by Amy

Belasco of the Congressional Research Service (Library of Congress), dated September 22, 2006.

... On Santa's Earnings and Real-Versus-Fake Tree Sales

My information on what a genuinely bearded shopping mall Santa Claus earns in a typical season (chapter 5) comes from the Amalgamated Order of Real Bearded Santas and the National Association of Professional Santas. In the article "With Santa's Return, a Happy Ending at Tysons" (Fredrick Kunkle, *Washington Post*, November 23, 2008), readers learn that the longtime Santa at Tysons Corner Center in Virginia, for one example, grosses about $30,000 a season.

My numbers on Christmas tree sales (in chapter 9) come from the National Christmas Tree Association (which promotes real trees) and the American Christmas Tree Association (which promotes artificial trees); there is no disinterested third party taking a census of American Christmas tree preferences.

... On Present-Day Frisco

The *Forbes* ranking in chapter 1 comes from "America's Fastest-Growing Suburbs" (via Forbes.com), by Matt Woolsey, July 16, 2007.

The average household income, ages, home prices, etc., come from the Frisco Economic Development Corporation, the U.S. Census, the Frisco Chamber of Commerce, the City of Frisco, the Collin (County) Central Appraisal District, the Denton (County) Appraisal District, the National Association of Realtors, and other demographic sources that you can take or leave. The bottom line: Frisco is young, educated, highly reproductive, loves to shop, and, even in a down economy, has plenty of closet space and disposable income.

My numbers on Frisco's sales taxes, total retail square footage, and property values, referenced in "Circuit City (Some Endings)," are from the Frisco EDC.

The reference in chapter 1 to the number of Texans who believe the Bible to be the literal word of God comes from a poll conducted by the nonpartisan Texas Lyceum in 2007, the results of which were described in "Texans Not Lock-Step Religious" by Christy Hoppe, *Dallas Morning News*, June 13, 2007.

My reference to the number of 2007–8 home foreclosures in Frisco and the Colony (in chapter 20) comes from an interactive map maintained by the *Dallas Morning News*'s website (www.dallasnews.com), which combines data from the *News*, Foreclosure Listing Service, the North Central Texas Council of Governments, and ESRI (Environmental Systems Research Institute).

... On Shopping, Retail Sales, and Frenzies

Until recently, my appetite for business articles about the retail industry was limitless. My leafy-green diet consisted mainly of the *New York Times, Washington Post, Wall Street Journal, USA Today*, and many localized dispatches by one Maria Halkias of the *Dallas Morning News*. Of the heaps of information I absorbed into my tale, some should be credited specifically. Let's do that now.

IN CHAPTER 6

> D'Innocenzio, Anne. "Holiday Shopping Season Off to Strong Start; Wal-Mart Is an Exception." Associated Press, November 25, 2006; and "Retailers Have a Good Start to the 2006 Holiday Season, but Will Momentum Continue?" Associated Press, November 27, 2006.

Merrick, Amy, and Kris Hudson. "Holiday Sales Off to Solid Start, but Wal-Mart Doesn't Share Cheer." *Wall Street Journal*, November 27, 2006.

Mui, Ylan Q. "Boot Camp Before the Blitz: Retailers Ready Their Workers for 'Black Friday,' the Post-Thanksgiving Rush." *Washington Post*, November 22, 2006.

IN CHAPTER 14

Barbaro, Michael. "In Shirt-Sleeve Holiday Season, Overcoats Linger on the Racks." *New York Times*, December 23, 2006.

Halkias, Maria. "Weekend Is the Final Call for Last-Minute Shoppers: 'Full Saturday and Sunday Before Christmas' Has Retailers Looking Toward Strong Finish." *Dallas Morning News*, December 22, 2006.

Herman, Charles. "Holiday Shopping: It's Now or Never." ABC News (via ABCNews.com), December 22, 2006.

IN CHAPTER 19

Issitt, Monica. "Going Great Guns: From Frisco's Farms to Its Future." *Frisco Style*, February 2008.

Ross, Wilbur. "The Psychological Impact Will Only Get Worse," from "We Ask: When Will the Pain Go Away?" *Newsweek*, June 16, 2008.

IN "CIRCUIT CITY (SOME ENDINGS)"

Lynch, David J. "Signs of a Growing Crisis: 'Relentless Flow' of Bad Economic News Suggests There's No Easy Way Out; 'Payback' for Debt-Fueled Growth?" *USA Today*, July 16, 2008.

Mardele, Susan. "What Recession? Frisco Holds Its Own." *Frisco Style*, May 2008.

Saitz, Greg. "Ho-ho-hum Holiday Season: Retail Sales Are Looking Pretty Bleak This Winter." *(New Jersey) Star-Ledger*, October 26, 2008.

Walker, Elaine, and Ina Paiva Cordle. "Sales Make Last-Minute Shopping a Buyer's Dream." *Miami Herald,* December 20, 2008.

... *Reheated Morsels*

In a few places I've repurposed a small bit of my newspaper work. In "The Gap (A Slide Show)," the passage about Sociable the cat's drowning on Christmas first appeared, in slightly different form, in "Woe, Woe, Woe, Merry Christmas," *Washington Post,* December 25, 2005. In chapter 14 and in "Circuit City (Some Endings)," a few observational quips about the foundering retail economy first appeared in "Pall on the Mall," *Washington Post,* November 29, 2008; and "This Is Dispirit of Christmas Present," *Washington Post,* December 24, 2008.

... *Oh, Pioneers*

My information about the forgotten graves discovered in September 2006 at the Dallas North Tollway construction site in Frisco (from chapter 14) relied initially on the following:

Hartzel, Tony. "Old Grave May Lie in Tollway's Path." *Dallas Morning News,* September 15, 2006.
——. "A 2nd Grave at Toll Site." *Dallas Morning News,* September 16, 2006.
Rathburn, Penny. "Looking Forward Uncovers the Past." *Frisco Enterprise*/Star Community Newspapers, October/November 2006.

From there I pursued more details on my own and was greatly aided by Ben Beckelman, who believes that the infant bones discovered at the site may be those of the daughter of his great-grandparents, John George and Fannie Kelsay, who named the baby Ireba; she died on December 14, 1902. I obtained further information from a working report delivered to the North Texas Tollway Authority in the summer of 2007, titled "Archeological Investigation at the Sonntag Family Cemetery, Collin County, Texas," prepared by AR Consultants, Inc. I quote from it in chapter 19.

In search of details about daily life and commerce in the area around the time the Kelsay baby died, I came upon a trove of children's letters to Santa Claus in several issues of the *Daily Courier* newspaper in McKinney, Texas. Letters that I quote in chapter 14 appeared in the *Courier* on December 21 and 22, 1902; and December 8 and 24, 1903.

For Further Reading

Christmas Histories, Origins, Culture

Barnett, James H. *The American Christmas: A Study in National Culture.* New York: Macmillan, 1954.

Elliot, Jock. *Inventing Christmas: How Our Holiday Came to Be.* New York: Harry N. Abrams, 2001.

Forbes, Bruce David. *Christmas: A Candid History.* Berkeley: University of California Press, 2007.

Highfield, Robert. *The Physics of Christmas: From the Aerodynamics of Reindeer to the Thermodynamics of Turkey.* Boston: Little, Brown, 1998.

Kelly, Joseph F. *The Origins of Christmas.* Collegeville, MN: Order of St. Benedict, 2004.

Mendelson, Lee. *"A Charlie Brown Christmas": The Making of a Tradition.* With reminiscences by Bill Melendez. Edited by Antonia Felix. New York: HarperCollins, 2005.

Miles, Clement A. *Christmas Customs and Traditions: Their History and Significance.* New York: Dover Publications, 1976. First published as *Christmas in Ritual and Tradition: Christian and Pagan* in 1912 in London by Fisher Unwin.

Rosen, Jody. *White Christmas: The Story of an American Song.* New York: Scribner's, 2002.

Skinner, Georja. *The Christmas House: How One Man's Dream Challenged the Way We Celebrate Christmas.* Novato, CA: New World Library, 2005.

Standiford, Les. *The Man Who Invented Christmas: How Charles Dickens's 'A Christmas Carol' Rescued His Career and Revived Our Holiday Spirits.* New York: Crown, 2008.

Waits, William B. *The Modern Christmas in America: A Cultural History of Gift Giving.* New York: New York University Press, 1993.

Christmas Problems

Flynn, Tom. *The Trouble with Christmas.* Amherst, NY: Prometheus Books, 1993.

Gibson, John. *The War on Christmas: How the Liberal Plot to Ban the Sacred Christian Holiday Is Worse Than You Thought.* New York: Sentinel/Penguin, 2005.

Law, Stephen. *The Xmas Files: The Philosophy of Christmas.* London: Weidenfeld & Nicolson/Orion Publishing Group, 2003.

McKibben, Bill. *Hundred Dollar Holiday: The Case for a More Joyful Christmas.* New York: Simon & Schuster, 1998.

Robinson, Jo, and Jean Coppock Staeheli. *Unplug the Christmas Machine: A Complete Guide to Putting Love and Joy Back into the Season.* New York: Quill/William Morrow & Co., 1991.

Classic Christmas Stories

Dickens, Charles. *A Christmas Carol and Other Christmas Writings.* With an introduction by Michael Slater. New York: Penguin Classics, 2003.

Dickens, Charles. *A Christmas Carol and Other Stories.* With an introduction by John Irving. New York: Random House/The Modern Library, 2001.

Moore, Clement C. (attributed). *The Night Before Christmas* (alt. *A Visit from St. Nicholas* or *'Twas the Night Before Christmas*). Illustrated by Mary Engelbreit. New York: HarperCollins, 2004.

Consumerismo!

Bongiorni, Sara. *A Year Without 'Made in China': One Family's True Life Adventure in the Global Economy.* New York: Wiley, 2007.

Cohen, Lizbeth. *A Consumer's Republic: The Politics of Mass Consumption in Postwar America.* New York: Alfred A. Knopf, 2003.

DeChant, Dell. *The Sacred Santa: Religious Dimensions of Consumer Culture.* Cleveland: Pilgrim Press, 2002.

DeGraaf, John, David Wann, and Thomas H. Naylor. *Affluenza: The All-Consuming Epidemic.* San Francisco: Berrett-Koehler, 2001.

Frank, Robert. *Richistan: A Journey Through the American Wealth Boom and the Lives of the New Rich.* New York: Crown, 2007.

Gangnier, Regenia. *The Insatiability of Human Wants: Economics and Aesthetics in Market Society.* Chicago: University of Chicago Press, 2000.

Lavin, Maud, ed. *The Business of Holidays.* New York: The Monacelli Press, 2004.

Levine, Madeline. *The Price of Privilege: How Parental Pressure and Material Advantage Are Creating a Generation of Disconnected and Unhappy Kids.* New York: HarperCollins, 2006.

Levitch, Timothy "Speed." Excerpt from *Speedology, Speed on New York on Speed.* New York: Context Books, 2002. Reprinted as "Wall Street: The Story of What Happened to Our Intimacy." In *The Outlaw Bible of American Essays.* Edited by Alan Kaufman. New York: Thunder's Mouth Press, 2006.

Mitchell, Stacy. *Big-Box Swindle: The True Cost of Mega-Retailers and the Fight for America's Independent Businesses*. Boston: Beacon Press, 2006.

Sosnik, Douglas B., Matthew J. Dowd, and Ron Fournier. *Applebee's America: How Successful Political, Business, and Religious Leaders Connect with the New American Community*. New York: Simon & Schuster, 2006.

Talen, Bill (writing as "Reverend Billy"). *What Would Jesus Buy? Fabulous Prayers in the Face of the Shopocalypse*. New York: Public Affairs, 2006.

Underhill, Paco. *Call of the Mall*. New York: Simon & Schuster, 2004.

———. *Why We Buy: The Science of Shopping*. New York: Simon & Schuster, 1999, revised 2009.

Whybrow, Peter C. *American Mania: When More Is Not Enough*. New York: W. W. Norton, 2005.

And Just a Little More Texas

Bainbridge, John. *The Super-Americans: A Picture of Life in the United States, as Brought into Focus, Bigger than Life, in the Land of the Millionaires — Texas*. New York: Doubleday, 1961.

Collin County Historical Preservation Group, Inc. *Historical Markers of Collin County, Texas*. McKinney, TX: Collin County Historical Preservation Group, 2005.

Graff, Harvey J. *The Dallas Myth: The Making and Unmaking of an American City*. Minneapolis: University of Minnesota Press, 2008.

Marcus, Stanley. *Minding the Store: A Memoir*. Boston: Little, Brown, 1974.

Marcus, Stanley. *Quest for the Best*. New York: Viking, 1979.

Rogers, John William. *The Lusty Texans of Dallas*. New York: E. P. Dutton, 1965.